Field Guide to the
Ladybirds
of Great Britain and Ireland

HELEN ROY AND PETER BROWN

ILLUSTRATED BY RICHARD LEWINGTON

BLOOMSBURY WILDLIFE

LONDON · OXFORD · NEW YORK · NEW DELHI · SYDNEY

Dedication

To David, Katy, Ella, Clare, Cameron and Jodie; and in memory of Wendy Young,
who inspired and nurtured Helen's love of wildlife.

BLOOMSBURY WILDLIFE
Bloomsbury Publishing Plc
50 Bedford Square, London, WC1B 3DP, UK

BLOOMSBURY, BLOOMSBURY WILDLIFE and the Diana logo are trademarks of
Bloomsbury Publishing Plc

First published in Great Britain 2018

A catalogue record for this book is available from the British Library

ISBN: HB: 978-1-4729-3567-0; PB: 978-1-4729-3568-7;
ePDF: 978-1-4729-3570-0; ePUB: 978-1-4729-3569-4

2 4 6 8 10 9 7 5 3 1

Typeset in Frutiger and Linotype Centennial by D & N Publishing, Baydon, Wiltshire
Printed and bound in China by RR Donnelley

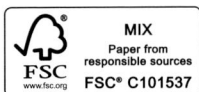

FSC
www.fsc.org

MIX
Paper from
responsible sources
FSC® C101537

To find out more about our authors and books visit www.bloomsbury.com
and sign up for our newsletters

Contents

Foreword

On August Bank Holiday 2017, instead of braving the traffic, I wandered up the garden and casually turned over a Sycamore leaf ... as you do. Beneath it, dwarfed by the sheltering spiders, was a barely visible amber hemisphere glinting in the dappled light. Through a hand lens, it became a minute beetle with a yellow horseshoe on its shining mahogany wing-cases – my first *Clitostethus arcuatus*, one of the tiniest and least recorded of British ladybirds. I think I cheered out loud. The honour of hosting such a scarce insect was out of all proportion to its diminutive size, so it's good to see that in these pages it has been awarded a common name, the Horseshoe Ladybird.

Most ladybirds are more familiar. Their bright colours, burnished wing-cases, pleasing, rounded shapes and usefulness as pest controllers have long endeared them to us. Their ambassador is the aphid-munching 7-spot Ladybird, whose scarlet wing-cases, freckled with black dots, endowed it with religious associations in the Middle Ages when the Virgin Mary was depicted wearing red to symbolise the blood of Christ. The seven spots represented her seven Joys and Sorrows – and so, bearing its badges of divine approval, it became known as Our Lady's Bird.

Not all the 47 British ladybirds have spots, but they are all described and illustrated in this ground-breaking identification guide, which is the perfect marriage of artistic excellence, deep knowledge and, dare I say it, of scientists' genuine affection. Alongside the familiar species – the 2-spot, the 7-spot and the new 'bird on the block, the invasive Harlequin – you will find the elusive set previously known only to the keenest of the coleopterati, the 'inconspicuous ladybirds'. Most are small, many are overlooked, and each is a call to arms for potential recorders to swish their nets through pastures, peer behind bark, riffle through leaf-litter and shake vegetation in the hope of discovering it. Is the Horseshoe Ladybird more widespread than we thought? Will the almost mythical False-spotted Ladybird give itself up? Ladybirds offer great scope for new discoveries on a national as well as a local level, and this guide will guarantee more dots on maps for the UK Ladybird Survey.

Ladybirds really do have it all. Among the 47 is a nice mix of common and rare, specialist and generalist species, enough to provide challenges for beginner or expert. They rival orchids and dragonflies as year-twitch subjects – surely the writing of the first ladybird quest diary is only a matter of time? But they offer deeper analysis too. How will the fortunes of the Harlequin and its predators play out? How will climate and landscape change affect our ladybirds?

I hope this book prompts more ladybird study. These loveable insects, which feature so often in children's literature and even have a publishing imprint named after them, have lives that seem harsh by human standards. Some species, including the familiar 7-spot Ladybird, are attacked by a parasitoid wasp whose larvae devour them from inside, Alien-style, before descending to pupate between the ladybird's legs. Many ladybird larvae are cannibalistic, and some species harbour male-killing bacteria whose Machiavellian manipulations were intensively researched by the late Michael Majerus, a champion of ladybird study. His baton has now passed to Helen Roy and Peter Brown, who have placed ladybirds firmly centre stage and, with this guide, shone the spotlight on these charismatic beetles. Beautifully illustrated by Richard Lewington, this book will enthuse and inspire a new generation of ladybird spotters (sorry!), and for me it will be an essential companion on future field trips.

Talking of which, Striped Ladybird would be good …

Brett Westwood, naturalist, author and broadcaster

Acknowledgements

We are exceptionally grateful to the thousands of people who have contributed ladybird records over many decades. Over the last 10 years the number of recorders has increased, and together we have been able to make considerable progress in understanding the life histories of these fascinating beetles. We also owe a debt of gratitude to many county Coleoptera recorders, other beetle recorders, Local Environmental Record Centre staff and natural history societies, for sharing their ladybird records with us. Ladybird recording in Ireland has been led by various people over the decades, with Roy Anderson, Gill Weyman and the National Biodiversity Data Centre playing major roles. We thank the Guernsey Biological Records Centre and the Jersey Biodiversity Centre for Channel Islands species lists.

We are grateful to Darren Mann and Amoret Spooner at the Oxford University Museum of Natural History, Max Barclay at the Natural History Museum, and Sami Karjalainen, for their generosity in allowing access to their wonderful collections of ladybirds. We are extremely appreciative of the invaluable advice on the taxonomy of Coccinellidae that Oldřich Nedvěd at the University of South Bohemia has provided.

We thank Katy Roper (Bloomsbury), who has shown considerable commitment to the production of this field guide, plus Hugh Brazier (copy-editor) and D & N Publishing (design). Martin Harvey reviewed the text, and we are extremely grateful for his thoughtful comments and advice. The following authors of the regional accounts kindly shared their local knowledge of ladybird ecology: Derek Bateson, Katie Berry, John van Breda, Richard Comont, Rachel Farrow, Trevor James, Paul Mabbott, Alex Pickwell, Don Stenhouse and Gill Weyman. We thank the photographers for their generosity in sharing their images, particularly Gilles San Martin.

The Biological Records Centre (BRC), which is co-funded by the Natural Environment Research Council through the Centre for Ecology & Hydrology (CEH) and the Joint Nature Conservation Committee (JNCC), hosts the UK Ladybird Survey and provides support in many ways including the production of maps and analysis of distribution trends. We are grateful to Jim Bacon, Martin Harvey, Jodey Peyton and Biren Rathod for developing and supporting the websites: www.ladybird-survey.org and www.coleoptera.org.uk. Many ecologists within the BRC have contributed their expertise, and we specifically thank Colin Harrower and Steph Rorke, who managed the Coccinellidae database and produced the maps; Charlie Outhwaite and Nick Isaac, who analysed the dataset to derive the distribution trends; and David Roy, who has assisted in extracting information from iRecord and the Coccinellidae database. Anglia Ruskin University supported Peter's contribution to the book, particularly through its sabbatical fund.

We thank the external communications teams at CEH and Anglia Ruskin University for their enthusiasm and support in promoting the UK Ladybird Survey so widely, especially Paulette Burns, Barnaby Smith and Jon Green. Indeed, there are many people within both CEH and Anglia Ruskin University who have enthusiastically supported our activities over the years, including Marc Botham, Trevor James, Chris Preston, Mark Hill, Richard Pywell, Rosie Hails, Mark Bailey, David Roy, Sheila Pankhurst, James Johnstone and Alvin Helden. Many Masters and PhD students have worked with us over the last decade: Marco Benucci, Katie Berry, Richard Comont, Rachel Farrow, Will Fincham, Claire Kessell, Gabriele Rondoni, Kate Titford, Sandra Viglášová and Trish Wells. Their research has greatly enhanced our understanding of many aspects of ladybird ecology.

The Field Studies Council, and specifically Rebecca Farley, have supported the UK Ladybird Survey over many years, including in the production of the field charts *Guide to Ladybirds of the British Isles* and *Guide to the Ladybird Larvae of the British Isles*, alongside the atlas *Ladybirds (Coccinellidae) of Britain and Ireland*.

Finally, we would like to thank two people who provided inspiration and joy to us in our early days of coordinating the UK Ladybird Survey: Mike Majerus and Bob Frost. We very much wish they were still able to join us in our entomological exploits.

Introduction

The stirring of insects from their overwintering sites is a tantalising sign of spring across Britain and Ireland. Every year the sight of 7-spot Ladybirds emerging from leaf litter to bask in the warm spring sun is met with delight. On cold days, they seem to disappear again, but then appear sporadically over the weeks leading into March and April.

Undoubtedly the bold colours and striking patterns of ladybirds are the main reason for the popularity of this insect group to so many people. The bright markings are warning colouration to indicate to predators that ladybirds are slightly toxic and bitter-tasting. There are some ladybirds, however – the so-called inconspicuous species – that are not brightly coloured. It is these hidden treasures, alongside their charismatic relatives, that we hope to introduce you to in this field guide. All the species have varied life histories and fascinating behaviours that contribute to their appeal. Some are predators and feed on pest insects, much to the delight of gardeners and farmers. Others feed on plants, and some graze on mildews that occur on plants. However, we still have so much to discover, and we hope that this guide will help you not only to get new insights into these seemingly well-known beetles, but also, through your own field observations, to add to what is known about them. As with so many other insect species, the distributions of ladybirds are shifting in response to climate change, and we look forward to you joining us in documenting the occurrence and unravelling the ecology of these much-loved creatures.

▲ 7-spot Ladybirds emerging from their overwintering sites, with a single Harlequin Ladybird.

Diversity of ladybirds

Ladybirds are beetles (insect order Coleoptera) and have their own family, the Coccinellidae. Worldwide there are about 6,000 species, making it a moderate-sized beetle family. There are 360 ladybird genera, with 45 of these occurring in Europe and 26 in Britain. Europe has approximately 260 species, of which 47 are regarded as resident in the British Isles. Britain and Ireland tend to have relatively low numbers of insect species compared to other European regions because of their separation from mainland Europe and the cooler climate.

The geographical scope of this book is Britain (i.e. England, Wales and Scotland, including offshore islands), Ireland (both Northern Ireland and the Republic of Ireland), the Channel Islands (Guernsey, Jersey and several other small islands) and the Isle of Man. For simplicity we will mostly refer just to 'Britain and Ireland' without specifying the Channel Islands and the Isle of Man, but it can be assumed that the associated text is relevant to these islands. Sometimes we will refer only to Britain, as not all species are found in Ireland.

The 47 ladybird species listed in the *Checklist of Beetles of the British Isles* (Duff 2018) are listed in the table on pages 8–10. While all of them occur in Britain, Ireland has 28 of the species, and in the Channel Islands 30 species have been recorded.

The family Coccinellidae is divided into subfamilies and tribes in which the genera and species are placed. However, the evolutionary relationships between different groupings of ladybirds are not yet fully resolved, and the subfamilies are considered to be polyphyletic, which means they are descended from more than one common ancestor. In Britain and Ireland there are six tribes – Coccidulini, Scymnini, Chilocorini, Ortaliini, Coccinellini and Epilachnini – within one subfamily, Coccinellinae. The Coccidulini and Scymnini are often referred to as the 'inconspicuous' species, and are small, indistinct and slightly hairy insects that would not be recognised as ladybirds by many people. Indeed, these inconspicuous ladybirds (alongside *Platynaspis luteorubra*) have not previously been given common names – but in the species accounts in this book we propose some names to bring them into line with their conspicuous counterparts. The remaining four tribes comprise mostly the large conspicuous ladybirds; they are not generally hairy, with the exception of the Epilachnini and Ortaliini (of which in Britain and Ireland the only species is the non-native *Rodolia cardinalis*). Of the 47 species of ladybird resident in Britain and Ireland, 27 are designated as conspicuous species.

▲ Two of Britain's inconspicuous species: *Coccidula scutellata* (left) and *Clitostethus arcuatus* (right).

Classification of the family Coccinellidae occurring in Britain and Ireland (Duff 2018, Roy *et al.* 2011, 2013)

Tribe	Scientific name with authority	Common name	Recorded in Britain?	Recorded in Ireland?	Recorded in Channel Islands?
Inconspicuous species					
COCCIDULINI Mulsant, 1846	*Coccidula rufa* (Herbst, 1783)	Red Marsh Ladybird	Yes	Yes	Yes
	Coccidula scutellata (Herbst, 1783)	Spotted Marsh Ladybird	Yes		
	Rhyzobius chrysomeloides (Herbst, 1792)	Round-keeled Rhyzobius	Yes		
	Rhyzobius forestieri (Mulsant, 1853)		Yes		
	Rhyzobius litura (Fabricius, 1787)	Pointed-keeled Rhyzobius	Yes	Yes	Yes
	Rhyzobius lophanthae (Blaisdell, 1892)	Red-headed Rhyzobius	Yes		
	Cryptolaemus montrouzieri Mulsant, 1853	Mealybug Destroyer	Yes		
SCYMNINI Mulsant, 1846	*Hyperaspis pseudopustulata* Mulsant, 1853	False-spotted Ladybird	Yes	Yes	Yes
	Clitostethus arcuatus (Rossi, 1794)	Horseshoe Ladybird	Yes		
	Nephus bisignatus (Boheman, 1850)	Two-spotted Nephus	Yes, but thought to be extinct		Yes
	Nephus limonii (Donisthorpe, 1903)		Yes		
	Nephus quadrimaculatus (Herbst, 1783)	Four-spotted Nephus	Yes		
	Nephus redtenbacheri (Mulsant, 1846)	Red-patched Nephus	Yes	Yes	
	Scymnus haemorrhoidalis Herbst, 1797	Red-rumped Scymnus	Yes		Yes
	Scymnus limbatus Stephens, 1832	Bordered Scymnus	Yes	Yes	
	Scymnus auritus Thunberg, 1795	Oak Scymnus	Yes	Yes	Yes
	Scymnus suturalis Thunberg, 1795	Pine Scymnus	Yes	Yes	
	Scymnus femoralis (Gyllenhal, 1827)	Heath Scymnus	Yes		
	Scymnus frontalis (Fabricius, 1787)	Angle-spotted Scymnus	Yes		Yes
	Scymnus interruptus (Goeze, 1777)	Red-flanked Scymnus	Yes		Yes
	Scymnus jakowlewi Weise, 1892		Yes		
	Scymnus nigrinus Kugelann, 1794	Black Scymnus	Yes	Yes	

Tribe	Scientific name with authority	Common name	Recorded in Britain?	Recorded in Ireland?	Recorded in Channel Islands?
SCYMNINI Mulsant, 1846 continued	Scymnus rubromaculatus (Goeze, 1777)		Yes		
	Scymnus schmidti Fürsch, 1958	Schmidt's Scymnus	Yes	Yes	Yes
	Stethorus pusillus (Herbst, 1797)	Dot Ladybird	Yes		Yes
Conspicuous species					
CHILOCORINI Mulsant, 1846	Chilocorus bipustulatus (Linnaeus, 1758)	Heather Ladybird	Yes	Yes	Yes
	Chilocorus renipustulatus (Scriba, 1791)	Kidney-spot Ladybird	Yes		Yes
	Exochomus quadripustulatus (Linnaeus, 1758)	Pine Ladybird	Yes		Yes
	Platynaspis luteorubra (Goeze, 1777)	Ant-nest Ladybird	Yes		Yes
ORTALIINI Mulsant, 1850	Rodolia cardinalis (Mulsant, 1850)	Vedalia Beetle	Yes		
COCCINELLINI Latreille, 1807	Halyzia sedecimguttata (Linnaeus, 1758)	Orange Ladybird	Yes	Yes	Yes
	Psyllobora vigintiduopunctata (Linnaeus, 1758)	22-spot Ladybird	Yes	Yes	Yes
	Vibidia duodecimguttata (Poda, 1761)	12-spot Ladybird	Yes	Yes	
	Anisosticta novemdecimpunctata (Linnaeus, 1758)	Water Ladybird	Yes	Yes	Yes
	Coccinula quattuordecimpustulata (Linnaeus, 1758)		Yes		
	Tytthaspis sedecimpunctata (Linnaeus, 1761)	16-spot Ladybird	Yes		Yes
	Hippodamia tredecimpunctata (Linnaeus, 1758)	13-spot Ladybird	Yes	Yes	Yes
	Hippodamia variegata (Goeze, 1777)	Adonis Ladybird	Yes		Yes
	Aphidecta obliterata (Linnaeus, 1758)	Larch Ladybird	Yes	Yes	
	Adalia bipunctata (Linnaeus, 1758)	2-spot Ladybird	Yes	Yes	Yes
	Adalia decempunctata (Linnaeus, 1758)	10-spot Ladybird	Yes	Yes	Yes
	Coccinella hieroglyphica Linnaeus, 1758	Hieroglyphic Ladybird	Yes	Yes	
	Coccinella magnifica Redtenbacher, 1843	Scarce 7-spot Ladybird	Yes		Yes
	Coccinella quinquepunctata Linnaeus, 1758	5-spot Ladybird	Yes		Yes

9

continued overleaf

Tribe	Scientific name with authority	Common name	Recorded in Britain?	Recorded in Ireland?	Recorded in Channel Islands?
COCCINELLINI Latreille, 1807 continued	Coccinella septempunctata Linnaeus, 1758	7-spot Ladybird	Yes	Yes	Yes
	Coccinella undecimpunctata Linnaeus, 1758	11-spot Ladybird	Yes	Yes	Yes
	Harmonia axyridis (Pallas, 1773)	Harlequin Ladybird	Yes	Yes	Yes
	Harmonia quadripunctata (Pontoppidan, 1763)	Cream-streaked Ladybird	Yes	Yes	
	Propylea quattuordecimpunctata (Linnaeus, 1758)	14-spot Ladybird	Yes	Yes	Yes
	Anatis ocellata (Linnaeus, 1758)	Eyed Ladybird	Yes	Yes	
	Myrrha octodecimguttata (Linnaeus, 1758)	18-spot Ladybird	Yes	Yes	
	Calvia quattuordecimguttata (Linnaeus, 1758)	Cream-spot Ladybird	Yes	Yes	
	Myzia oblongoguttata (Linnaeus, 1758)	Striped Ladybird	Yes	Yes	Yes
EPILACHNINI Mulsant, 1846	Henosepilachna argus (Geoffroy in Fourcroy, 1762)	Bryony Ladybird	Yes		
	Subcoccinella vigintiquattuorpunctata (Linnaeus, 1758)	24-spot Ladybird	Yes	Yes	Yes

Notes to table: scientific names are in two parts, the genus and species. It is conventional to italicise these and to add the surname of the person (or people) who originally described and published the particular scientific name – the so-called 'authority'. The date of publication of the description is often given after the authority. If the genus name changes after the original description then the original authority's name is given in brackets. The common names for the conspicuous ladybirds (except *Platynaspis luteorubra*) have been in use for many decades, but here we propose simple and descriptive common names for the inconspicuous ladybirds too.

Alternative names have been used for some genera and species in the past, including: *Henosepilachna* (= *Epilachna*); *Adonia* (= *Hippodamia*); *Tytthaspis* (= *Micraspis*); *Myzia* (= *Neomysia*); *Psyllobora* (= *Thea*); and *Coccinella magnifica* (= *C. distincta*).

A comprehensive checklist of Coleoptera (including synonyms) has recently been published (Duff, 2018). It recognises the division of the Coccinellidae into tribes, across which there are 55 species, including one (*Nephus bisignatus*) that is considered extinct in these islands. A further seven species, for the most part, have only a few records and there is limited or no evidence of their establishment. In addition, *Nephus limonii* is added as a reinstated species; it is thought to be a cryptic sibling species of *N. redtenbacheri*, and although these two species are difficult to separate on the basis of morphological characteristics, *N. limonii* seems to be strongly associated with sea-lavender *Limonium*. With the exception of *Rhyzobius forestieri* and *Vibidia duodecimguttata* (see *Potential new species*, pages 152–153), these additional species are not considered in this field guide beyond mention in this table (where they are marked with a grey background).

▲ a. Typical ladybird from tribe Coccidulini: *Coccidula rufa*; b. typical ladybird from tribe Scymnini: *Scymnus suturalis*; c. typical ladybird from tribe Chilocorini: Kidney-spot Ladybird; d. typical ladybird from tribe Coccinellini: 2-spot Ladybird; e. typical ladybird from tribe Epilachnini: Bryony Ladybird.

Characteristic features

Adult ladybirds are small to medium-sized beetles (1.3–10mm in length). Male and female adult ladybirds look similar, although males are often smaller than females. Both sexes have oval, oblong oval or hemispherical body shapes with convex upper surfaces and flat lower surfaces, resulting in the domed shape.

Looking at the head it is possible to see the large compound eyes, and close to the inner front edge or below the eye, more or less clubbed antennae, which are usually eleven-segmented (but can have as few as seven segments). The mouthparts are made up of large, strong mandibles, and just behind these are four-segmented maxillary palps with an axe-shaped end segment; a labium (lower lip), divided into the pre-labium and post-labium, which are connected by a membrane; three-segmented labial palps; and the labrum, or upper lip. The maxillary and labial palps are used by the ladybird to give sensory information about food before ingesting it.

Behind the head is a hard plate called the pronotum, which covers the thorax. The pattern on the pronotum can be extremely instructive for identifying a species. The head can be partly withdrawn under the pronotum, which is wider than the head and broader than long, with slight anterior extensions at the margins.

Ladybirds have short legs, which they can retract into depressions under the body. The tarsi (end sections of the legs) bear two claws and are usually four-segmented, but it is very difficult to see the third segment and so the tarsi appear three-segmented (and species in the genus *Nephus* have only three segments).

The abdomen has ten segments, but from below only five, six or seven of these are obvious. The first abdominal segment often has a humped central area. The elytra (wing-cases) cover the top of the abdomen and when not in flight the membranous wings are folded away completely between the elytra and the abdomen. Some species are very varied in their elytral colour patterns. Such so-called polymorphism can cause confusion when identifying adult ladybirds, because individuals within one species can look very different.

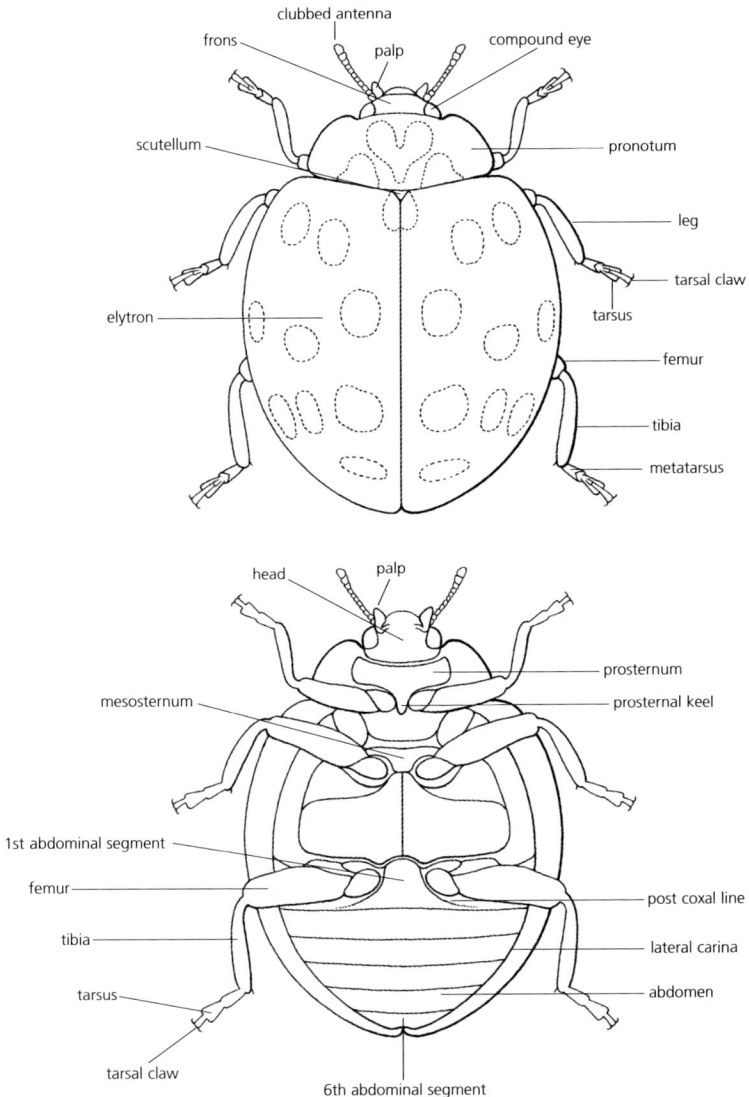

clubbed antenna
frons
palp
compound eye
scutellum
pronotum
leg
tarsal claw
elytron
tarsus
femur
tibia
metatarsus

head
palp
prosternum
mesosternum
prosternal keel
1st abdominal segment
femur
post coxal line
tibia
lateral carina
tarsus
abdomen
tarsal claw
6th abdominal segment

Summary of some major features of genera

Genus	Size	Body shape	Antennae	Hairy?	Main food
Coccidula	Small	Elongate, slightly convex	Long	Yes	Aphids
Rhyzobius	Small	Elongate, moderately convex	Long	Yes	Aphids
Hyperaspis	Small to medium	Oval, strongly convex	Short	No	Mealybugs and aphids
Clitostethus	Minute	Oval, strongly convex	Long	Yes	Whiteflies
Nephus	Minute	Oval, moderately convex	Short	Yes	Mealybugs and aphids
Scymnus	Minute to small	Oval, convex	Medium	Yes	Aphids and scale insects
Stethorus	Minute	Oval, moderately convex	Short	Yes	Mites
Chilocorus	Medium	Round, strongly convex	Short	No	Scale insects
Exochomus	Medium	Broadly oval, slightly convex	Short	No	Scale insects
Platynaspis	Small	Oval, slightly convex	Short	Yes	Aphids
Halyzia	Large	Oval, slightly convex	Long	No	Mildew
Psyllobora	Medium	Oval, moderately convex	Long	No	Mildew
Anisosticta	Medium	Elongate, flat	Medium	No	Aphids
Tytthaspis	Small	Oval, strongly convex	Medium	No	Mildew
Hippodamia	Medium to large	Elongate, slightly convex	Long	No	Aphids
Aphidecta	Medium	Oval, slightly convex	Long	No	Aphids
Adalia	Medium	Oval, slightly to moderately convex	Medium	No	Aphids
Coccinella	Medium to large	Oval to broadly oval, moderately to highly convex	Medium	No	Aphids
Harmonia	Large	Oval, slightly to highly convex	Medium	No	Aphids
Propylea	Medium	Oval, moderately convex	Medium	No	Aphids
Anatis	Large	Oval, slightly convex	Long	No	Aphids
Myrrha	Medium	Oval, slightly convex	Long	No	Aphids
Calvia	Medium to large	Oval, convex	Medium	No	Aphids
Myzia	Large	Oval, moderately convex	Long	No	Aphids
Henosepilachna	Large	Oval, convex	Long	No	Plants
Subcoccinella	Medium	Oval, strongly convex	Medium	Yes	Plants

▼ Types of ladybird antennae.

Long
(e.g. *Rhyzobius*)

Medium
(e.g. *Coccinella*)

Short
(e.g. *Chilocorus*)

Life history

Ladybirds have a system of complete metamorphosis as they develop; in other words, they are holometabolous. This also applies to other beetles (Coleoptera), and to other insect groups such as butterflies and moths (Lepidoptera) and flies (Diptera). These insects therefore have juveniles (larvae) that look very unlike the adults into which they will develop. More ancient insect groups, such as grasshoppers and crickets (Orthoptera), dragonflies (Odonata) and true bugs (Hemiptera), have incomplete metamorphosis (hemimetabolous). These groups have juveniles (nymphs) that look rather like miniature versions of the adult forms (though lacking wings). Only the holometabolous insects, such as ladybirds, have a pupal stage, during which the full-grown larva pupates and metamorphoses into the adult form.

All ladybirds therefore go through four stages of development: egg, larva, pupa and adult. However, whereas the life cycles of many conspicuous ladybirds are well studied, this is not the case for the inconspicuous ladybirds. The duration of the life cycle is dependent on temperature, but aphid-feeding species tend to develop much faster than scale-insect-feeding species, which in turn seem to develop faster than the mildew- and plant-feeding species. All have broadly similar life cycles, although the variation in some of the life-history details adds to the fascination of studying these beetles. In this section we provide an overview of the general life cycle of ladybirds in Britain and Ireland, with a focus on the conspicuous predatory ladybirds.

▲ Life stages of 7-spot Ladybird: egg, larva, pupa and adult.

EGGS

Most ladybird species lay small, smooth, yellow-orange elongated ovoid eggs in the spring (and again later in the summer for those species that have more than one generation). Depending on temperature, the eggs take about 4–5 days to hatch. Unlike butterfly eggs, ladybird eggs are mostly rather similar in appearance and it is not usually possible to determine the species from

▲ A typical cluster of yellow, ovoid ladybird eggs.
◄ Adonis Ladybird laying eggs.

the egg. However, eggs of some ladybirds are noticeably different: for example, the Chilocorini lay eggs that are less elongated than those of Coccinellini, and eggs of some Scymnini have a short thread-like protuberance. Eggs are usually laid upright, though the Chilocorini and Scymnini are exceptions and tend to lay eggs on their sides. Ladybird eggs are often laid in neat tight clusters, which distinguishes them from those of the Chrysomelidae (leaf beetles) that do not lay their eggs in an ordered, upright position but in slightly shambolic clusters.

Ladybirds lay eggs in batches of 1–100. For example, the 2-spot usually lays clusters of 10–20 eggs, and over 1,500 eggs may be laid in the female's lifetime. Clusters of Harlequin eggs are often quite large, comprising 20–50 eggs. Food consumption by the female ladybird (in terms of both quality and quantity) affects the number of eggs laid, and the size of the hatching larvae.

The female adult ladybird uses many cues to determine the suitability of her chosen egg-laying site; for the predatory species, this is often the underside of a leaf or stalk infested with aphids or scale insects (coccids). She will only lay her eggs near prey that have the potential to sustain her offspring, so will be deterred by excess shed skins of the prey, or by honeydew (the sticky excretion of aphids), which could signify a colony about to crash. The female ladybird can also detect the presence of other ladybird larvae, which leave a trail of 'footprint' chemicals behind. These larvae present a danger, as they could potentially consume the eggs or newly hatched larvae of the female. The eggs take about five days to hatch, and the emerging larvae are highly cannibalistic: a larva that feeds on a late-hatching sibling is more likely to survive than one that must catch a prey item soon after hatching.

Ladybird eggs have a relatively strong outer shell (chorion), compared to those of other beetles, and this is needed as eggs are often laid in exposed positions. The shell contains a wax layer and the eggs are chemically defended against predation, to a greater or lesser extent. Eggs of the Cream-spot are particularly well defended by waxy chemical compounds and are unpalatable to other ladybirds such as the Harlequin.

LARVAE

On hatching, ladybird larvae are tiny but rapidly increase in size, shedding their skin to pass through four stages (instars) before pupating. Ladybird larvae tend to have the same food preferences as adults of the same species and feed voraciously through their four larval instars. A late-instar aphid-feeding (aphidophagous) ladybird larva will consume as many aphids as an adult ladybird (up to 80 aphids per day). The larval stages progress over several weeks, and it is not uncommon through June to find mainly larvae and few adults, as the overwintered adults die off after reproduction. Early (first and second) instar larvae are very small and difficult to identify to species level, though with careful observation using a hand lens they may be identified to tribe or genus. Late (third and fourth) instar larvae are relatively easy to identify to species with the exception of 2-spot and 10-spot Ladybirds, which are tricky to distinguish from one another. The shade of the larva may vary depending on temperature, age and feeding. A larva that has recently shed its skin is often soft and pale in colour. The body darkens somewhat as the exoskeleton hardens, but as the larva feeds its new skin may become stretched, and it may again look paler.

▲ Emerging Harlequin Ladybird larva.

PUPAE

Just prior to pupation, the larva will adhere to a surface using a structure at the end of the abdomen called the anal cremaster. The substrate is often a leaf or other plant structure, but some ladybirds pupate on hard surfaces such as walls. The larva will then hunch over – in a stage known as the pre-pupa – and the larval skin will split for a final time. In species of Coccinellini the larval skin is shed completely, leaving only a remnant at the base. Species of the tribe Chilocorini pupate within their final larval skin, which is not fully ruptured until the adult beetle emerges. This is also the case for species in the tribes Scymnini and Epilachnini, but their larval skin is left slightly more open around the pupa than for the Chilocorini. Some pupae are therefore relatively distinctive and species identification is possible, particularly for the conspicuous ladybirds. The colouration and sculptured appendages of the pupae all assist identification, but the remnants of larval skin (the exuvia) can also be used. This is particularly the case for the Epilachnini, because the larvae have profuse protuberances that are clearly visible even after the skin is shed during pupation. As for other life stages, the colour of the ladybird pupa is variable, and may be affected by age, temperature and humidity. At the onset of pupation the pupa may appear pale, but it darkens throughout the pupal stage, although this only happens to a limited extent in warm, dry weather.

◀ Typical pupa from tribe Chilocorini: Pine Ladybird.

◀ Typical pupa from tribe Coccinellini: 7-spot Ladybird.

► Typical pupa from tribe Epilachnini: 24-spot Ladybird.

Pupae are vulnerable to predation as they cannot run away. However, they are not defenceless, and can flick up at the unattached end to startle a potential predator or parasitoid. Like other life stages, pupae also contain distasteful chemicals that provide further defence.

Although the pupa is sessile and so appears to be dormant, the processes of metamorphosis taking place during the pupal stage are rapid and result in a remarkable transformation. Many organs and tissues of the larva are broken down and used to form the new organs (wings, genitalia, mouthparts, and so on) of the adult ladybird.

ADULTS

The adult beetle emerges from the pupa after 5–7 days – a process termed eclosion. The newly emerged adult will sit on the pupal case for some time to expand the membranous flight wings and elytra and to allow the exoskeleton to harden. At this stage the elytra appear translucent and pale-coloured (yellow) because the pigments are laid down after eclosion. Therefore, the characteristic colours of a ladybird can take several days to fully develop (and indeed the beetle will continue to darken somewhat over its lifetime). However, the colour patterns on the pronotum are fully developed on emergence, and can be extremely helpful for identification. The shape of the adult ladybird may also change somewhat during its first few hours, being rather more elongate when it initially emerges.

PHENOLOGY

The study of the timing of recurring natural events, such as bud burst or spring emergence of species, is called phenology. The timing of life-history events of different species has become especially important in recent times because there have been shifts in response to climate change. Phenology has highlighted that for some species key events, such as emergence from winter, are occurring earlier than was previously the case. The timing of the different stages of the life cycles of ladybirds varies between species and differs each year, depending on weather conditions. Similarly, each species of ladybird moves into its overwintering habitat at different times, with the Harlequin Ladybird being one of the last to do so. Little is known about the timing of life-history events for the inconspicuous ladybirds.

Not all species are likely to shift their life-history events in parallel. There are concerns that if shifts for ladybirds and their food occur asynchronously over time with climate warming, then these predators and their prey may become mismatched. As an example, aphids have a short generation time (about a week), and can develop even at low temperatures. Therefore, they are expected to respond strongly and rapidly to environmental change. In contrast, ladybirds have a comparatively long generation time (a matter of months), and so their response to environmental change will differ from that of aphids. Some ecologists are concerned that there could be an increase in pest outbreaks with this mismatch in predator activity, but the complexity of community interactions makes this difficult to predict.

COMMUNITY INTERACTIONS

The interactions between ladybirds and other species has captivated ecologists for decades. Such community interactions are important for evaluating the biological control potential of species complexes. So, many studies have focused on the way in which predators and parasites of aphids interact and how this may affect the control of pest insects. In some cases, it has been shown that some combinations of natural enemies of aphids are synergistic (that is, their combined effects are greater than either one or another acting alone), and in other cases they have negative consequences for pest control and are termed antagonistic. The varying life histories of the different species, and specifically variation in the timing of life-history events, can help with predictions about the ways in which natural enemies will behave when living in close association.

The foraging behaviour of predatory insects is complex and influenced by many factors, including the architecture of the plant but also chemical signals being emitted by the plants, the aphids, the other natural enemies and combinations of all of these. Unravelling the network of interactions among these species is challenging, but will undoubtedly progress over the coming years as new techniques emerge to support such studies.

The Harlequin Ladybird provides an excellent example of the ways in which an understanding of community ecology provides critical insights particularly for assessing the effects of a new arrival on existing species assemblages. This non-native species is known to interact with many other natural enemies of aphids, mostly in negative ways, as a predator and strong competitor of the other species. It is now the dominant predator in many habitats across its invaded range. However, it is unclear how the dominance of the Harlequin and subsequent changes to community interactions will alter the dynamics of aphid populations and ultimately crop production. It is possible that Harlequin Ladybirds will more effectively control pest insects than all the other natural enemies put together, but it is perhaps more likely that some of the nuances of the pest control system will be lost with the demise of some of the previously common species. There are well-founded concerns that the loss of ecological interactions, as a species declines in abundance, will result in the loss of ecological function even though the species remains. In other words, the loss of ecological interactions is of equal concern to the extinction of species.

DORMANCY

Winter in Britain and Ireland presents potentially hostile conditions for ladybirds, with limited food availability for predatory species. They therefore move into a state of dormancy (overwintering) to escape the worst of the weather – effectively, they hibernate, although this is a term used only for endotherms (warm-blooded species). The dormancy period lasts from five to eight months, often starting in September and ending in March. However, this varies by species. Harlequin Ladybirds, for example, tend to remain active until later in the year (late October), and while Pine Ladybirds emerge from overwintering early (February), 14-spot Ladybirds typically emerge late (April). The start and end of overwintering is triggered by day length and temperature.

There are two main types of dormancy for ladybirds in Britain and Ireland: quiescence, often a short-term state from which the individual can emerge at any time should conditions become favourable again; and diapause, which often lasts for a longer period and through which some aspect of physiological development occurs. Quiescence is controlled by external factors such as temperature, while diapause is controlled by internal factors. Both of these dormancy types may be employed during a winter. For example, a 7-spot Ladybird will enter a period of diapause in late autumn; later in the winter, as spring approaches, it will enter quiescence.

▶ Harlequin Ladybirds overwintering.

In the latter state it can readily 'wake up' if the weather warms, and shut down again if it becomes cooler. In this way the ladybird can temporarily come out of dormancy to have a drink or even find food. The non-predatory ladybirds also overwinter, but since food is more likely to be available to them, they may become active in warmer periods of the winter. In Britain and Ireland, all ladybirds overwinter as adults. The only known exception is the Orange Ladybird, which may (rarely) overwinter as a pupa.

In hotter climates (in southern Europe, for example) some ladybirds have a dormancy period at the hottest time of year, in summer. Ladybirds may also escape the heat of summer by moving to higher altitudes. Perhaps this will also become the case in Britain and Ireland as the climate continues to change.

In a state of dormancy, ladybirds can tolerate very cold temperatures. So, while very harsh conditions may cause mortality, it seems that fluctuating conditions are more damaging to them because valuable energy reserves are used in the warm periods and depleted more rapidly than would be the case if temperatures were consistently low. This appears to be an increasing problem as climate change leads to more unpredictable weather. Mortality of ladybirds can be high in winter, but varies year on year.

Ladybirds need cool, sheltered places in which to overwinter, and a variety of habitats are used. Depending on the species (see *Species accounts*, pages 68–145, for details), ladybirds may overwinter in leaf litter, hollow plant stems, grassy tussocks or the soil. Some species use crevices in tree bark, or overwinter among conifer needles. A couple of species tend to move to buildings. Many ladybirds form groups (aggregations) for overwintering, and sometimes these are of mixed species. Chemical signals (pheromones) may be used to alert ladybirds to an overwintering site. Aggregating in winter may give the dormant ladybirds protection, or may help them in terms of seeking a mate when they disperse in spring to breed.

MATING

On emergence from winter dormancy, ladybirds will begin the process of reproduction. For most species this will mean dispersing from their overwintering habitat to find food and then a mate, but a few species will mate with other individuals within their winter group before feeding. Some species will need to mate many times to maintain egg laying, but for others one mate can provide sufficient viable sperm for many months. Indeed, 2-spot Ladybirds can lay fertile eggs for many months following a single mating. Sperm can be stored over winter, and this has been shown to be the case for Kidney-spot Ladybirds.

▲ Mating 7-spot Ladybirds.

The female ladybird wandering around a plant while carrying a male is a familiar sighting throughout the spring and summer months. While she continues to forage and feed, the male will often be seen ferociously rocking from side to side. It is thought this is happening during the release of sperm, packaged within a spermatophore, but it also causes the dislodgement of sperm from other males.

Ladybirds are highly promiscuous. The benefit to males of multiple mating is the potential to fertilise many eggs. For females there are also many advantages. It has been shown that mating has a stimulatory effect on the rate of oviposition by females, with egg laying increasing over a few days after mating and then declining unless there is further mating. Additionally, a female who mates with many males is more likely to have encountered a genetically fit male. Multiple mating also leads to the potential for sperm competition, whereby the sperm from two or more males compete to fertilise the egg, although the evidence of exactly how this might work is unclear.

FEEDING

Many species of ladybird are considered to be beneficial because they feed on sap-sucking insects such as aphids, scale insects and adelgids (small bugs also called woolly conifer aphids). Gardeners are generally delighted to see ladybirds because they provide some level of pest control. Indeed, with some ladybirds consuming in excess of 60 aphids per day, the potential for control of pest insects is huge. However, in many cases ladybirds have been unsuccessful when introduced deliberately as biological control agents, particularly when the target pest is an aphid species. This is simply because aphids are so prolific and multiply so rapidly, reaching numbers that even the hungriest of ladybirds cannot consume. The life cycle of an aphid can be completed within a week, and so aphids can go through many generations over the spring and summer months, whereas most species of ladybird only have one generation per year. There has been more success with control of scale insects by ladybirds, primarily because the durations of their life cycles are more matched than those of aphids and ladybirds.

▶ 7-spot
Ladybird
feeding on
aphids.

▶ 22-spot
Ladybird
feeding on
mildew.

▶ 24-spot
Ladybird
feeding on
vegetation.

Most of the ladybirds found in Britain and Ireland principally eat aphids or scale insects, but there are also non-predatory species that feed on plants (24-spot Ladybird and Bryony Ladybird) or mildew (16-spot Ladybird, 22-spot Ladybird and Orange Ladybird). Many species of ladybird have diverse diets, with the Harlequin Ladybird perhaps being one of the most extreme in this regard, feeding primarily on aphids but also on many other insects and even on soft fruits.

Not all foods will provide the necessary nutrition to support reproduction in ladybirds. So, while some foods may be acceptable for building up energy reserves to see the adult through the winter months, there will be others that are critical for the reproductive success of a female ladybird (so-called 'essential food'). Intriguingly, there are cases where a food can be toxic to one species and essential to another – the Mealy Plum Aphid, for example, is an essential prey for 7-spot Ladybirds but is toxic to 2-spot Ladybirds. However, in general there are many different species of essential prey for each ladybird, and so lots to choose from.

Some species of ladybird are known to be extremely specialised with respect to prey choice. For example, *Stethorus pusillus* feeds on mites within the genus *Phyllacotes*, while the three species of *Nephus* and *Hyperaspis pseudopustulata* are reported to feed primarily on scale insects within the genus *Pseudococcus*, and *Scymnus auritus*, an oak specialist, is thought to feed exclusively on the aphid *Phylloxera glabra*.

The feeding ecology of ladybirds is undoubtedly complex, and there is still much to reveal about food choice. Many reports are based on small laboratory studies in which the ladybirds were exposed to prey species that they may not encounter in the wild. Other inferences of predator–prey interactions have been based on the co-occurrence of ladybirds and likely pest insects that are assumed to be prey on the basis of proximity. However, feeding behaviour can vary from one geographic region to another and throughout the year.

The cannibalistic behaviour of ladybirds has been well studied, and occurs mainly in the larval stages. It is critical that a newly hatched larva feeds as quickly as possible. For the non-predatory species this can be straightforward: the plant and mildew feeders are likely to be surrounded by easily accessible food. However, a tiny predatory larva might be the same size or smaller than the prey it finds itself surrounded by. It is quite remarkable to watch a small predatory ladybird larva riding on the back of an aphid as it feeds voraciously. Some larvae don't wait to find food but instead turn on their siblings hatching around them. This can improve the survival chances and reduce the development time of a ladybird larva. Some larvae hatch within an egg cluster in which only females are viable; their mother carries a bacterium within her cells that can only be transmitted from mother to daughter, and so to maximise transmission all the male embryos are killed by the bacterium. These so-called 'male-killing bacteria' are intriguing and, although it seems a macabre life-history strategy, the hatching females benefit enormously from the consumption of a dead brother.

Cannibalism is not restricted to the newly hatched ladybirds. Larvae at all stages will consume one another, and pupating individuals are particularly vulnerable to attack from neighbouring larvae. Harlequin Ladybirds can reach very high numbers on deciduous trees during the summer months, and as they exhaust the supply of aphids they will turn on one another (and other natural enemies of the aphids, including other ladybird species). It is not unusual to see late-instar larvae attacking pupae from the underside, where they can resist the defensive flicking motion. Less commonly, adult ladybirds may be seen engaging in cannibalism.

Mouthparts of ladybirds vary, being adapted to the specific food type that dominates the diet of each species. The inner side of the mandibles of ladybirds can be divided into an incisor and a molar region. The predatory ladybirds have two large sharp tooth-like structures on the incisor and molar regions, whereas the herbivorous ladybirds have many blunt and small teeth along the mandibles that are specialised for biting and grinding up plant material. The mildew-feeding species have rows of small teeth for gathering fungal spores.

Ladybirds use various senses to detect their food, and ecologists are increasingly recognising the importance of chemical signalling within and between species. It is fascinating to think of the way in which predatory ladybirds perceive the world as various chemical plumes being emitted from plants and from their prey. Chemical signals emitted by plants are known to change when the plant is under attack from pest insects, and this is a way in which the plant can call in 'bodyguards' – that is, the ladybirds and other natural enemies of plant pests that are attracted to the chemicals produced by damaged plants.

The foraging behaviour of ladybirds is strongly influenced by the structure of the plants that are host either to them or to their food (pest insects or mildew). Smooth leaves can be tricky for small larvae to forage over, and hairy leaves may also present problems. However, foraging beetles can usually find a way to overcome such physical barriers: for example, on shiny leaves they will focus movement around the veins. The foraging behaviour of different ladybird species varies considerably, and it can be rewarding to observe and compare species. Harlequin and 7-spot Ladybirds share many features – they are similar-sized, voracious, generalist predators. However, while their larvae are relatively well matched as predators, adult 7-spots are slower than Harlequin Ladybirds in recognising and capturing prey; a 7-spot Ladybird adult can bumble across a leaf infested with aphids yet seemingly failing to attack. This is one of the reasons why Harlequin Ladybirds have been more successful in controlling aphids in biological control programmes than 7-spot Ladybirds. However, both species will feed on non-pest insects, including other ladybirds, and this could have adverse ecological consequences by contributing to species declines and ultimately changes in the composition of insect communities.

Larval and adult ladybirds often consume the same types of food, but the predatory ladybirds differ in their feeding strategies at different stages of development. The larvae mainly inject enzymes into their prey and then suck up the partially digested contents. It is

▲ Cannibalism by siblings.

not unusual to see an aphid wandering around a plant in a seemingly dishevelled state – its insides having been partially consumed by a ladybird larva, leaving just the husk of the aphid which can still (quite remarkably) walk around. In contrast, aphid-feeding adult ladybirds, while also injecting enzymes into the prey, generally consume the entire aphid, often leaving only the legs and antennae. There is considerable variation between species in feeding strategy: *Clitostethus arcuatus* feeds mainly on whitefly eggs, which it pierces and then sucks out the contents; *Platynaspis luteorubra* has perforated mandibles and sucks the fluid directly from its aphid prey.

MOVEMENT, FLIGHT, MIGRATION AND DISPERSAL

The larval and adult stages of ladybirds are highly mobile. Ladybirds can walk quite rapidly and negotiate all kinds of surfaces. Larvae are obviously restricted to movement by walking but can still travel reasonable distances, moving from one plant to another in search of food. Movement is temperature-dependent, but on a warm sunny day larvae may be seen moving rapidly and with incredible agility, navigating the most demanding plant architectures. For example, Scarce 7-spot larvae and adults move around Scots Pine trees, managing to cope with the challenges of oozing resins, fine needles and aggressive wood ants.

Perhaps one of the most notable features of an adult ladybird is the pair of hard forewings (elytra) that for many species are brightly coloured. Hidden beneath these are the delicate membranous hindwings that are revealed in flight. It is intriguing to watch adult ladybirds launch themselves into the air. They will often walk to the edge of a leaf, the tip of a blade of grass for example, and quite slowly raise their elytra out and forward so that the lower edges face to the front and effectively reduce air resistance. However, the elytra also have an important role in providing lift and steering in flight.

The hindwings consist of membranous panels attached to a framework of veins. On emergence from its pupa, an adult ladybird stretches out the hindwings and pumps the veins to provide the necessary structural support for these aerofoils. During one wingbeat the aerodynamic forces continually shift, and these small insects can accommodate these pressures and cope with unexpected impacts such as colliding with a rain drop. For the most part the front vein is reinforced and partly hidden by the hard forewing, but the tip, appropriately called the deformable tip, is less rigid and crumples on impact, but regains shape rapidly.

◄ Harlequin Ladybird launching into flight.

The wings, unlike those of vertebrates, do not contain muscles and instead gain the energy necessary for flight from the thoracic muscles. The veins and areas of thickened membrane work as levers, amplifying small movements to enable the upstroke and downstroke necessary for flight. These membranous wings are neatly folded away after landing. They have a number of weak points, which enables this sophisticated action to take place. However, it is the folded nature of the membranous wings that results in some problems for ladybirds launching into flight. Ladybirds cannot leap into the air but need their leg movements to coordinate with the first stroke of the wings, and so the hindwings must be meticulously unfolded first – this accounts for the slowness of the onset of flight – a pause while the ladybird prepares the hindwings. Majerus (1994) described this 'as though the ladybird, like an airline pilot, is going through a checklist to ensure against mishap, before applying the power'.

Ladybirds can fly considerable distances. The flight of some large species (including Harlequin and 7-spot Ladybirds) has been explored using information from an innovative upward-facing radar. These ladybirds can fly high – often 150–500m high, and up to 1,100m high over long distances – enabling them to move throughout the landscape in search of prey, mates and suitable habitats. Indeed, ladybirds have been reported far out to sea when they alight on cruise ships, and they can cross the English Channel in the air, with optimal weather conditions. The long-distance movement of ladybirds can lead to the arrival of new species not previously encountered.

During the summer months, ladybirds will mostly take short flights to travel between patches within habitats, alighting when they get signals that suggest food is likely to be abundant. How exactly the ladybirds make decisions about suitable habitats is unclear. There have been many studies on short-distance prey recognition by predatory ladybirds, particularly those that feed on aphids, but understanding of long-range prey location is lacking. Hungry ladybirds are more attracted to yellowish-green than to other colours, and it is quite possible that this attraction is adaptive, because prey insects are often situated on the pale growing shoots of plants. Once on a plant, foraging ladybirds follow the leaf veins and edges – again, highly likely to bring them into contact with aphids. Ladybirds also use chemical cues to home in on their prey, but of course the prey can fight back, and do so in a variety of ways. Aphids, for example, excrete a substance called alarm pheromone, which warns nearby aphids to flee. Some ladybirds are known to perceive and orientate towards alarm pheromone but will sometimes stop foraging to remove droplets of this slightly sticky substance if they come into direct contact with it.

▲ 7-spot Ladybird larva with aphid exuding alarm pheromone.

PREDATORS AND PARASITES

The conspicuous ladybirds are aposematic; that is, the bold contrasting patterns represent warning colouration. In addition, when disturbed, ladybirds pull their legs into depressions on the abdomen and exude a strong-smelling yellow fluid (reflex blood). Aposematism and reflex bleeding are effective deterrents against predators, but still some birds, spiders and even other insects (including other species of ladybird) will attack and consume ladybirds.

▲ Scarce 7-spot Ladybird with wood ant.

Some ants and aphids have a mutualistic relationship in which the ants benefit from the sweet honeydew produced by the aphids and the aphids in return gain protection against predators and parasites. Therefore, ants are largely intolerant of ladybirds. Confrontations rarely result in the death of the ladybird, but more commonly the exclusion of the ladybird from the colony of aphids being protected by the ant. There are exceptions, however, and a few species of ladybird in Britain live alongside ants. The Scarce 7-spot Ladybird and *Platynaspis luteorubra* are considered to be essentially 'ant-loving' (myrmecophilous) and live in the company of ants that seem intolerant of other ladybird species. The way in which this relationship has evolved is unclear, but undoubtedly these myrmecophiles have filled niches that many other aphid predators cannot thrive within and so, to some extent, are freed from competition.

A number of parasites use ladybirds as hosts. Some are extremely specialised and have intriguing life cycles while others are opportunistic and have a broad host range. The latter include some of the pathogenic fungi, such as *Beauveria bassiana*, that cause disease in a wide range of insects. The specialised parasites include flies (Diptera) and wasps (Hymenoptera), which are termed parasitoids because they develop within the host, eventually killing it.

◄ Harlequin Ladybird pre-pupa with parasitic fly.

The most common parasitic flies of ladybirds are within the genus *Phalacrotophora* (in the family Phoridae). *Phalacrotophora fasciata* and *P. berolinensis* commonly attack ladybirds in Britain and Ireland. They are superficially very similar in appearance, with 2mm long, squat bodies, which are pale yellow-brown and characteristically hunched. Phorids usually lay their eggs between the legs of pre-pupal ladybird larvae or newly formed pupae. It only takes a few hours for the eggs to hatch, and then they burrow into the host to complete their larval development, before emerging from the ladybird host to pupate. Some of the larger species of ladybird, such as the 7-spot and Eyed Ladybirds, can contain more than six individual phorids, whereas smaller species, such as the 2-spot Ladybird, usually only have three or four parasites emerging. There have been records of phorids from the non-native Harlequin Ladybird, but seemingly at lower prevalence than from ladybird species native to Britain and Ireland. This is certainly worthy of further study, and the collection of ladybird pupae to assess emergence (either of an adult ladybird or of a parasite) is straightforward (see Roy *et al.* 2013). *Medina separata*, a tachinid fly, has also been recorded from ladybirds in Britain. This parasite looks very like a housefly but is 5–6mm in length, with a slender, hairy abdomen.

The braconid wasp *Dinocampus coccinellae* is the most prevalent hymenopteran parasite of ladybirds. This parasitoid is well studied but still there is much to reveal about its life history. The adult is about 4mm long and dark-bodied with iridescent green-black eyes. As it approaches an adult ladybird, it thrusts its abdomen under the body and stabs at the host. The parasitoid egg is laid within the body of the ladybird and hatches within about five days; the parasitoid larva develops within the host, completing three stages (instars). Only one parasitoid can develop within the ladybird, so if more than one larva hatches within the host then one will destroy the other. The developing parasitoid acquires nutrients from the ladybird haemolymph and from specialised feeding cells produced by the parasitoid within the ladybird. The developing larva emerges from the abdomen of the ladybird, which it partially paralyses but does not kill, and spins a cocoon under the legs of the ladybird. It is this cocoon that is the most conspicuous stage of this parasitoid, and it is commonly observed. The ladybird host is attached to the cocoon and essentially acts as a bodyguard to the parasitoid. The adult parasitoid emerges after about a week and in most cases the ladybird host dies of starvation. *D. coccinellae* can have several generations in a year but in Britain and Ireland it usually has just one generation. It overwinters as a larva within the host.

D. coccinellae has been recorded from many species of ladybird but some are more suitable than others. It can be particularly prevalent in large ladybirds such as the 7-spot, whereas small ladybirds seem less suitable. The Harlequin Ladybird is more resistant to the parasitoid than are native ladybirds, but this is predicted to change over time as the wasp adapts to this novel host. A number of other wasps have been recorded parasitising ladybirds in Britain, including small chalcid wasps of the genera *Oomyzus*, *Aprostocetus* and *Homalotylus*.

▶ 7-spot Ladybird with cocoon of parasitoid *Dinocampus coccinellae.*

Studying and recording ladybirds

There are many reasons why ladybirds are a rewarding group of insects to study. A number of species can be found commonly within gardens or other urban environments, but there are some species that require venturing further afield to particular habitats. Many species can be identified in the field, both as adults and as late-instar larvae, without specialist equipment. The UK Ladybird Survey welcomes records from everyone, and a smartphone app, iRecord Ladybirds, provides a convenient method for recorders to submit sightings. Alternatively, the general iRecord app includes an option to upload a photograph, which is particularly useful for subsequent verification (see *Identification and sending in records*, page 33).

As with many species groups, some ladybirds are habitat generalists, while others are more specialist. Some species are more tolerant of wetter and cooler conditions, so within their habitat may be found across Britain and Ireland, while others are more sensitive to climate and therefore have a more restricted distribution. In general, ladybirds prefer warmer and drier conditions, so the highest species richness tends to occur in southeastern England. Scotland and Ireland lack several species that are common and widespread in England and Wales. However, a good range of ladybird species may be found in most parts of Britain and Ireland. To maximise the number of species observed, several different habitat types need to be searched. Within these, a range of plant species may be sampled to maximise the species count. The undersides of leaves are a good place to begin searching, and in winter, leaf litter and bark crevices can be fruitful.

Within a site, plants with aphids are more likely to yield ladybirds, so will be worth investigating. The same applies to scale insects – rather strange insects that may sometimes be spotted on tree bark as bumps or hardened shell-like structures. Except in the case of two species – the Scarce 7-spot Ladybird and *Platynaspis luteorubra* – which are ant-loving (myrmecophilous), ladybirds and ants tend to be in conflict (see *Predators and parasites*, pages 26–27) so the presence of ants may mean fewer ladybirds.

WEATHER AND TIME

The best conditions for surveys are warm, dry and sunny days without too much wind. Ideally, the temperature should be above 14°C, although if it is sunny a cooler temperature may still yield good results, and even on rainy days ladybirds can be found with increased search effort. Windy conditions can make tree beating (and to a lesser extent sweep netting) a tricky task. Within a site, things to take into account when sampling ladybirds include aspect and shade, with sunny places likely to be more fruitful.

◀ Aphids on Reed.

Both larval and adult ladybirds are diurnal, and daytime surveys are recommended, particularly between the hours of 10am and 4pm, when ladybirds are at their most active. Adult ladybirds are more readily seen from March to September (the main active months) but the season can be extended if the weather is favourable. Overwintering habitats may be sampled from October.

EQUIPMENT

Ladybird surveys may be done without any equipment (beyond a notebook or smartphone for documenting your observations or records) – searching vegetation by eye can be a successful way of spotting ladybirds. However, this will tend to limit records to the larger and more brightly coloured species, and especially those that are active on deciduous trees (the leaves of which may make the insects easier to spot). There are a few pieces of equipment that are likely to make ladybird surveys more varied and rewarding, including a ×10 hand lens and a sweep net or beating tray.

Sweep net

A sweep net is used for sampling terrestrial insects in grasses, Nettles and other low herbaceous vegetation, and is an effective tool for finding ladybirds. The net comprises a strong canvas bag (of pale colour) on a metal frame. Nets are often collapsible for easier transport, and if low weight and size are particularly important (for example if taking the net when travelling), compact lightweight sweep nets with telescopic handles are available.

The net should be swept from side to side repeatedly through the vegetation, quite low to the ground. The number of sweeps carried out before the catch is checked is up to the recorder and may depend on how many insects are active. However, five to ten sweeps at a time may be about right. Too many sweeps may result in a lot of plant debris and insects

▶ Sweep netting.

gathering in the net, making it tricky to see the ladybirds. It is important to check through the debris at the bottom of a net after sweeping in search of the small inconspicuous ladybirds.

Methods can be standardised by sweeping for a set length of time, or keeping to a standard number of sweeps, over a known area of the habitat. This will allow comparisons if doing repeat surveys. Sweep netting is not effective in very low vegetation (less than about 15cm), which may need to be searched by eye, and will not capture insects from the soil surface. It should only be attempted in dry vegetation – the net gets messy if it is wet, and the invertebrates can easily be damaged.

Sweep nets are available from entomological suppliers and may be fairly expensive, but a good-quality net should last for many years. Butterfly nets are not generally suitable for sampling ladybirds, as they are more delicate and liable to tear in prickly low vegetation. Home-made sweep nets may be constructed out of canvas material, thick wire and wooden dowelling (for instructions see Roy *et al.* 2013). After sampling, the insects should be released back into the vegetation, unharmed.

Beating tray (or umbrella)
A beating tray is used for sampling in trees (either deciduous or coniferous) and shrubs. The tray essentially comprises a pale rectangular canvas stretched over a wooden frame. It folds away for compactness. The tray is placed under a tree branch and the branch is tapped several times with a stout stick (or broom handle), so that the insects fall onto the tray. It is important not to tap the branch too lightly (especially at first), as once alerted the insects may grip onto the vegetation. Again, standardised methods can be adopted by beating a set number of branches a specific number of times within a habitat.

There are a few alternatives to a beating tray. One option is simply to use a piece of pale-coloured material, placed on the ground, with branches beaten above it using a stick. A second option is to use an upturned umbrella in place of the beating tray. The umbrella should ideally be unpatterned and white (or at least a pale colour) so that the insects are more visible. This method was regularly used with great success by our late friend Bob Frost, a keen naturalist and ladybird recorder.

Sweep nets can also be used for tree beating (and are easier to use in windy conditions than a beating tray, which can become unwieldy). Holding a sweep net under the branches of trees or bushes and then hitting the branches firmly can yield ladybirds that are otherwise hidden. So if you are choosing either a sweep net or a beating tray, then the former is more versatile.

The ladybirds are usually still for a few seconds or more after they land on the tray or umbrella, but in warm weather they may fly off fairly quickly. If this becomes problematic, then beating into a sweep net is an option to consider. As with sweep netting, after sampling, the insects should be released back into the trees, unharmed.

Hand lens/microscope
Most of the conspicuous ladybirds, and some of the inconspicuous ones, can be identified in the field with a hand lens or by eye. A hand lens (×10 works best for us) is certainly a useful addition. Lenses (or loupes as some call them) are very variable in quality and it may be worth investing a bit more in a good-quality lens from an entomological supplier.

Some inconspicuous species are easier to identify under a microscope; indeed, for some species it may be essential to do this to ensure correct identification (separation of some *Rhyzobius* and *Scymnus* species, for example). It can be helpful to put the ladybird gently into a clear plastic bag so you can manipulate its position under the optics. In warm weather, ladybirds can be cooled in the refrigerator prior to examination to reduce activity (though they tend to warm up and become active again very quickly).

▲ Beating of low shrubs with flood debris.

▲ Tree beating.

◀ Field survey equipment: ×10 hand lens and small collection pot.

Small pots

Carrying a few small (15–50ml) clear plastic containers with lids is a good idea when sampling ladybirds. These may be especially useful in summer, when the catch may be higher and the insects more active. The insects can then be kept for a little while in order to identify or photograph them. Avoid adding any plant material to the pot, because this creates moisture, which can be detrimental to the ladybird.

▲ Everyone can get involved in recording ladybirds.

REARING LADYBIRDS

Ladybirds may be reared at home, and there is more information on this in Roy *et al.* (2013). A common problem when doing so is the frequent need to find aphids as food. Various alternative diets may be used instead of aphids to sustain ladybirds, and there is a good recipe in Roy *et al.* (2013) for an artificial diet. A small piece of apple, changed daily, will also contain enough water and sugar to keep the ladybird going for a few days, before release. However, many species of predatory ladybirds need aphids or scale insects as a source of food to supply all the necessary nutrition for successful breeding. Aphids can be found in reasonable numbers on Lime and Sycamore trees or Nettles; aphids from the latter are a particularly good source of nutrition for many species.

IDENTIFICATION AND SENDING IN RECORDS

Many people can assist in the identification of ladybirds, and often a photograph is sufficient to determine the species of a conspicuous ladybird (see section on *Websites, apps and social media*, page 154). In addition to the iRecord and iRecord Ladybirds apps, the UK Ladybird Survey has an e-mail address (ladybird-survey@ceh.ac.uk) and online recording forms (www.ladybird-survey.org/irecord.aspx), for submitting records. The organisers of the UK Ladybird Survey are always delighted to hear of ladybird observations from far and wide, and are happy to provide feedback on identification. There are also many county beetle recorders who are willing to respond to enquiries. Recently, social media has provided a useful means of corresponding with others regarding species identification, or indeed sharing ladybird sightings. The UK Ladybird Survey uses Twitter (@UKLadybirds) for these purposes.

▶ Using a mobile phone for ladybird recording.

Ladybird habitats

Ladybirds can be found in nearly all terrestrial habitats. Since Britain and Ireland have a range of generalist and specialist species, visiting several different habitat types will increase your chances of seeing the less ubiquitous ladybirds.

Grassland

Grasslands and meadows of various types are often good habitats for a range of ladybirds. Grassland specialists include the 16-spot and 24-spot, as well as the 22-spot, which may often be found on mildewed Hogweed in grassy margins. Unlike most ladybirds, these three species are not predatory. *Rhyzobius litura* and the 14-spot are aphid feeders that are common in grassland, while 11-spot and Adonis Ladybirds are typically less common predators in dry grassland. The 7-spot, 2-spot and Harlequin are generalists that may also be found in grassland habitats. Sweep netting is the best technique for sampling grassland ladybirds, especially as many of the grassland species are small and may occur close to the ground. Some, such as the 16-spot, are also camouflaged. For these reasons such ladybirds are often rather difficult to spot by eye.

Heathland

Heathland can be a rich habitat for ladybirds. Some, such as the Heather and Hieroglyphic Ladybirds, are specialists of the heathland habitat. Heather may be sampled using a sweep net. The Scarce 7-spot (which is always found close to wood ants) can be common in southern heathlands. Other ladybird species of coniferous trees and deciduous trees such as birch can occur in heathland habitats. Various ladybirds may be found in Gorse, so this is also worth beating. Indeed, some species use Gorse as an overwintering habitat and can be seen there even in the depths of winter.

Arable fields

Ladybirds are known for their role in controlling sap-sucking crop pests such as aphids, and may be sampled from a wide range of arable crops, especially in summer. Some species, such as the 7-spot, often feed on early-season aphids (for example on Nettle) before moving into arable crops later on, perhaps in late spring or early summer. If attempting ladybird sampling in arable fields, stick to field edges so as not to damage the crop. The grassy field margins themselves may also provide good habitats for a range of species (see *Grassland*, above).

Coastal

Several ladybird species are often quite coastal in their distributions and may be found on vegetation in dune systems and other habitats close to the sea. These include the 11-spot, Adonis and 24-spot. These species (perhaps especially the Adonis, which is a very common species in warm and dry areas of mainland Europe) are quite tolerant of dry conditions. The very rare 13-spot Ladybird sometimes also occurs in coastal areas (particularly the south coast of England), probably because individuals are migrating (or are blown in) from mainland Europe, where the species is more common. It is interesting to note that the Adonis Ladybird appears to have been expanding into new regions in recent years, presumably a consequence

▶ Grassland.

▶ Heathland.

▶ Arable
fields.

▲ Coastal landscape.

of our warming climate. However, the 11-spot Ladybird is becoming less common and correspondingly has a retracting distribution range. Observations of these species in the coming years will reveal whether these patterns are long-term trends.

Deciduous woodland and hedgerows

A range of common ladybirds can be found on deciduous trees. Aphid abundance, and therefore ladybird abundance, is often lower on trees in semi-natural woodland than on

◄ Deciduous woodland.

urban trees. This is thought to be a consequence of a number of factors including improved host-plant quality in urban habitats and so-called 'urban heat-island effects' whereby urban areas are warmer than rural localities and so insects such as aphids are able to reproduce more rapidly. The main ladybird species likely to be found on deciduous tree foliage are Harlequin, Cream-spot, 10-spot, 2-spot, Orange and 14-spot. Pine, and especially Kidney-spot, ladybirds are often seen on the bark of deciduous trees (with Kidney-spots favouring Ash and sallow). A combination of searching by eye and tree beating should be used to maximise the number of species recorded on deciduous trees.

Many deciduous tree species will host ladybirds, but Field Maple, Sycamore, Lime and birch can be particularly rewarding. Oaks and Beech tend to have lower numbers of ladybirds. Since the former are renowned as being very rich in insect life, this may seem surprising, but oaks have good chemical defences and aphids do not tend to be abundant on them, hence the relatively low number of ladybirds. However, 10-spot Ladybirds can be abundant on oak, as well as on Hawthorn, in hedgerows. Oaks, especially saplings, are sometimes rather prone to mildew infestations, and these mildewed oak leaves are worth searching for 22-spot Ladybirds. Ivy is typically not rich in ladybirds, but there are a couple of small species that are worth searching for (perhaps by gentle beating): *Nephus quadrimaculatus* and *Clitostethus arcuatus*. These species are relatively rare but are probably under-recorded, and *N. quadrimaculatus* in particular can sometimes be found in large numbers in Ivy.

Hedgerows are worth beating gently and may yield the same ladybird species as deciduous trees. Indeed, *Rhyzobius litura* is commonly found through such sampling, alongside 14-spot and 10-spot Ladybirds.

Coniferous woodland

Coniferous woodland is a rich habitat for several ladybird species, including *Scymnus suturalis*, Eyed, Cream-streaked, 18-spot, Striped, Larch and Pine Ladybirds (though the Pine Ladybird is also common on deciduous trees). Scots Pine tends to be especially rich in ladybirds, but non-native pines, Douglas Fir, larch, spruce and other conifers are also worth beating. Some ladybird species, notably Striped and Cream-streaked, specialise on pines and the aphid species that are found there.

Conifers are also used by some generalist ladybirds, for example the 7-spot, especially for overwintering. Mature pines tend to yield more of the specialist conifer ladybirds, but young trees may also be rich, especially for the generalists.

▶ Coniferous woodland.

The interiors of conifer woodlands, particularly plantations, are often dark and lacking in understorey. Additionally, there are likely to be few branches within reach, so sampling within the woodland can be unrewarding. Sampling trees at woodland edges, including along rides, is likely to be much more effective.

Urban

Many ladybird species are common in urban habitats such as gardens, parks and churchyards. Most of the generalist species may be found here, and often in higher abundance than in semi-natural habitats, as urban localities commonly have high aphid abundance – as previously explained. Microclimate is also important, and the urban heat-island effect may contribute to high ladybird numbers.

Deciduous trees in urban localities, especially Lime, may yield a substantial number of ladybird species, including Harlequin, Cream-spot, 10-spot, 2-spot, Orange, 14-spot and Pine Ladybirds. Horse-chestnut can be worth searching for 10-spot, Pine and other ladybirds. Within its very restricted range, the Bryony Ladybird may be common on White Bryony in urban locations.

Weedy areas (for example, in grass verges and on waste ground) may have lots of aphids and provide habitat for many of the grassland ladybird species and generalists mentioned above. See also the *Grassland* and *Deciduous woodland and hedgerows* sections, above.

Wetlands including reedbeds

The Water Ladybird is the main conspicuous ladybird to specialise in wetland habitats. This species may be found close to water on Reed, Reedmace and other wetland plants. At times aphids can be abundant here, attracting the Water Ladybird as well as more generalist species (see *Grassland*, page 34). Wet grasslands (for example river floodplains) may also be home to this species. Our rarest conspicuous ladybird, the 13-spot, is also typically found in wetland habitats, including marshes, fens and riverbanks. Two of the inconspicuous species are also wetland specialists: the common species *Coccidula rufa*, and the less common *C. scutellata*. Sampling is best done by careful sweep netting. Additionally, during the winter the leaf sheaths of Reed and Reedmace may be searched carefully by eye for overwintering ladybirds.

◀ Urban habitat: allotments.

38

▶ Wetland reedbed.

River shingle

The 5-spot Ladybird is a specialist of shingle banks beside fast-flowing rivers. Within its very restricted range, in such habitats the 5-spot may be quite abundant, foraging on sparse weedy vegetation. However, this rare ladybird is largely restricted to the Spey Valley area of Scotland, and to Wales (mainly close to rivers in a central band running from the English border to the west coast). The 5-spot is at the edge of its range in Britain, and occupies a wider range of habitats in continental Europe. Searching for the 5-spot by eye on sparse weedy plants (such as dock) growing in among the shingle may be the most rewarding (and least destructive) method of sampling. These ladybirds are often found in places very close to the edge of the water, seemingly a rather precarious habitat; presumably the river shingle is desirable in being relatively competitor-free. However, if there are Gorse bushes, or other plants, close by, these may be worth beating.

Other ladybirds that may be found in this habitat (but often further from the water, on grassy or weedy banks), include the 7-spot, 11-spot, 24-spot, 22-spot, 14-spot and *Rhyzobius litura*.

▶ River shingle bank.

Regional guides

This section comprises contributions provided by various ladybird recorders around Britain and Ireland. It is not intended to be a comprehensive guide to recording ladybirds in all of our regions; rather, we hope that it provides a snapshot of the diversity of habitats and sites potentially rich in ladybird species in a selection of regions in which there is active ladybird recording. Please note that some of the sites mentioned may require permission for access. Additionally, at some sites the sampling of ladybirds (using sweep netting or tree beating techniques, for example) may require special permission in advance.

Dorset

Dorset's climate and diversity of habitats make it a superb county for finding ladybirds. Although best known for the heathlands that dominate parts of the east of the county, Dorset also has swathes of rich grassland, woodland and unspoilt coastline.

Around the busy resorts of Poole and Bournemouth, the heathlands have been encroached on by urbanisation but still provide extensive habitats for ladybirds. Heathland specialists such as the Heather Ladybird can be found throughout the county. High numbers of Striped Ladybirds have been found at Arne (SY9787), an RSPB reserve bordering Poole harbour with extensive lowland heath and Scots Pine that also supports many of the other conifer specialists. The adjacent heaths at Studland (SZ0382) and Godlingston (SZ0182) are hugely biodiverse owing to their unique range of habitats, with coastal dune heath, wetland and damp woodland, and so support a variety of ladybirds, ranging from the Eyed Ladybird, Pine Ladybird and Scarce 7-spot Ladybird to *Coccidula rufa*.

Away from the heaths of the east, Dorset gives way to the rugged coastline of the Jurassic Coast, pockets of ancient woodland and extensive downland to the west and north. At Fontmell Down (ST8818), 7-spot, 14-spot, 2-spot and 24-spot Ladybirds are all abundant along the paths. The hills of Hod (ST8510) and Hambledon (ST8412), both near Blandford, are also worth visiting to look for grassland specialists.

◀ Durlston Head, Dorset.

Rich meadows provide excellent habitat for grassland ladybird species at Durlston National Nature Reserve (SZ0277), where visitors are rewarded with a superb display of wildflowers and invertebrate life in the spring and summer months.

John van Breda

Isle of Wight

The Isle of Wight has a mild southern climate and a rich variety of habitats suitable for many species of ladybird. There are many places to search, from the coastal cliffs on the south coast to the inlets on the north coast and the chalk grassland belt undulating across the centre. The Isle of Wight was one of the first places to report extremely high numbers of the Harlequin Ladybird, a species that is now thriving there.

Newtown has a mosaic of habitats extending over a small area that can be accessed easily from the National Trust car park (SZ4290). The scattered conifers around the car park provide a good place to look for 18-spot Ladybirds, while 7-spot and 14-spot Ladybirds are abundant. A large field extends to the marshes and provides an excellent habitat for 24-spot and 16-spot Ladybirds.

Ventnor (SZ5678) on the south coast has habitats characteristic of dramatic coastal erosion. This benefits the Glanville Fritillary butterfly, and also provides warm spots for 11-spot and Adonis Ladybirds.

Helen Roy

London and the Thames basin

While predominantly urban and suburban, London and the Thames basin has many habitats suitable for ladybirds, including heaths and grasslands.

Until the arrival of the Harlequin Ladybird, two of the most common ladybirds in urban areas in and around London were the 2-spot and Pine Ladybirds, which have now become relatively rare. These and other species (10-spot, Cream-spot, 14-spot and 7-spot) are found in domestic gardens, parks and cemeteries, especially where there are mixes of plant types and mature trees. Irregularly disturbed soil with ruderal plant cover can occur anywhere, and often attracts Adonis Ladybirds. Street trees, especially Lime and Sycamore, are important homes for common species.

▶ Newtown, Isle of Wight.

◀ Box Hill, Surrey.

Mixed habitats such as woods in grassland, for example at Alexandra Park (TQ2990) and Bostall Wood (TQ4777), have diverse ladybird communities. These include species such as the Orange Ladybird, which is occasionally very abundant on Beech (e.g. at Epping Forest TQ4399) and Ash (e.g. at Epsom TQ1860) in semi-natural woodland. Derelict buildings in urban areas often support ant colonies with populations of *Platynaspis luteorubra*.

Pine specialists (Striped, Eyed and 18-spot Ladybirds) are found in West Heath, Hampstead (TQ2686) and the streets of Dagenham, as well as in plantations in Kent and Surrey. Native Scots Pines and other conifers on heath, especially in Surrey, are home to the Heather Ladybird and various inconspicuous ladybirds. Larch plantations are not common in the region but the Larch Ladybird is occasionally found on other conifers, for instance at Hampstead and Box Hill (TQ1852). The Cream-streaked Ladybird is found on pines in Hampstead but also on a variety of small trees in unexpected places. The Kidney-spot Ladybird conventionally aggregates on sallows (e.g. at Rushey Mead TL3810) but is also found on urban trees including apple.

The Water Ladybird is broadly distributed on waterside rushes, for example at Dagenham (TQ4984). The 11-spot Ladybird is found along rivers, although not as commonly as at the coast, where it often occurs among 7-spot Ladybirds (e.g. at Harwich TM2432).

Molesey (TQ1468) was the discovery site of the Bryony Ladybird, which has become established in southwest London. The other herbivorous ladybird, the 24-spot, is abundant on acid grasslands such as Chingford Plain (TQ3895), where it is often accompanied by the 16-spot. 22-spot Ladybirds are generally found on mildewed umbellifers, especially in hedgerows in the suburbs.

Heaths and acid grasslands, especially with pine, are very productive places to look for ladybirds, including various *Scymnus* species and other inconspicuous ladybirds. Example sites include Oxshott (TQ1461), Wisley (TQ0758), Esher (TQ1262), Wimbledon (TQ2372), Chobham (SU9963) and Box Hill (TQ1751). The Scarce 7-spot Ladybird can be found in these habitats in or near wood ant colonies, especially at Esher Common.

Old stone walls in parkland with good growths of Ivy support *Stethorus pusillus* and *Nephus quadrimaculatus* – for example at Nonsuch Hall (TQ2363).

Paul Mabbott

Oxfordshire

Oxfordshire has the rolling chalk hills of the Cotswolds and Chilterns to the west and south, and a range of sandstones, mudstones and clays across the rest of the county. This mixed underlying geology makes for a diverse botanical assemblage, which in turn supports a wide range of ladybird species.

▶ Harcourt
Arboretum,
Oxfordshire.

Adonis and 11-spot Ladybirds, which are more commonly seen in coastal areas, can be found in car parks and road verges, where vegetation is sparse and the bare ground is warm, particularly in the gravel pits around Abingdon such as the Cothill Fen and Radley Lakes complexes (the Earth Trust's Thrupp Lake, SU5297, is a good starting place). The mixed low vegetation at these sites is also good for Kidney-spot, 16-spot, 24-spot and 2-spot Ladybirds, alongside some of the scarcer inconspicuous species including *Platynaspis luteorubra*, *Rhyzobius lophanthae*, *Scymnus interruptus*, *S. auritus*, and, at the water's edge, *Coccidula rufa*.

One of the best sites in the county for ladybirds of coniferous trees is Harcourt Arboretum, outside Nuneham Courtenay (SU5598). The site has a variety of mature exotic conifers that support good populations of Eyed, Larch and Cream-streaked Ladybirds, while Pine and 18-spot Ladybirds, *Rhyzobius chrysomeloides* and *R. lophanthae* are also present. These species are also present throughout Oxford itself on conifers, particularly in churchyards. The Bryony Ladybird has an outlier population in Oxfordshire a few miles down the road at Benson (SU6191).

One of our scarcest ladybirds, *Clitostethus arcuatus*, historically had a population hotspot around Godstow Abbey (SP4809), between Oxford city and Wytham Woods. Recent searches have failed to find the species here, but it has been found at Shotover Hill (SP5606).

Richard Comont

Hertfordshire

Hertfordshire lies on the northern flanks of the London Basin, with the chalk ridge of the Chiltern Hills along its northwestern flank. As such, its natural landscapes and habitats range from chalk grassland through clay and gravel woodlands to alluvial and aquatic habitats along river valleys that generally flow south into the Thames. The chalk ridge forms the natural watershed with the East Anglian Fen Basin to the northeast, so there are outposts of fenland along its northern edge.

Forty-two of the native British ladybird species have been recorded from Hertfordshire, making it seem a rich part of the country for the group. However, some species have not been seen for some time, especially species restricted to heathland, such as the Heather Ladybird and Hieroglyphic Ladybird, or nationally rare species like the 13-spot Ladybird.

Of those that still occur in the county, as with most areas, many are the common, widespread species found in most rough grassland, hedgerows, scrubby places and gardens, although some, such as the 2-spot Ladybird, have recently declined, probably owing to competition with the invasive Harlequin Ladybird, which has spread everywhere across the county since its arrival in 2004. Nevertheless, species such as the 16-spot Ladybird, the 24-spot

◀ Broxbourne Wood, Hertfordshire.

Ladybird and the 22-spot Ladybird are still frequent in rough grassland across the county. Hertfordshire's woodlands (e.g. Broxbourne Woods complex, TL3207, and Northaw Great Wood, TL2704) and wooded commons (e.g. Berkhamsted Common, SP9911, Chorleywood Common, TQ0396, and Nomansland Common, TL1712) and parklands are home to a good range of species, including the Cream-spot Ladybird, 10-spot Ladybird, Orange Ladybird (much increased over the last 30 years), Pine Ladybird (not restricted to pines) and Kidney-spot Ladybird (especially on Ash and willow trees).

The county has had conifer plantations for a long time, ranging from old parkland stands through to modern forestry, so many conifer specialists are resident, such as the Eyed Ladybird, 18-spot Ladybird, Larch Ladybird, Cream-streaked Ladybird and *Scymnus suturalis*. Its damp habitats, also, have their special species. Prime among these include the Water Ladybird, which remains frequent along Hertfordshire's rivers. Wet meadows and fens hold *Coccidula rufa*, as well as its rarer relative *C. scutellata*, which has occurred recently at King's Meads (TL3413). These river valley wetlands, and some gravel pits, also have species such as *Scymnus limbatus* (found recently at Waterford Heath, TL3115) and *S. haemorrhoidalis*. Dry chalky grassland or open gravel pits are home to a wide range of scarcer species, such as the Adonis Ladybird and *Scymnus* species, *S. frontalis* and *S. femoralis* in particular. One (private) gravel pit has recently produced the nationally scarce ladybird, *Platynaspis luteorubra*.

Trevor James

Worcestershire

A largely lowland county with several major river valleys running through it, Worcestershire also includes the southern edge of the Wyre Forest, and the eastern edge of the Malvern Hills. The Malvern Hills (SO7747) in particular are under-surveyed. The scarce *Nephus quadrimaculatus* is common on Ivy around Great Malvern and the Malvern Hills, and the area is probably worth examining for other inconspicuous ladybirds.

Nearby, the flood meadows of the Rivers Severn and Avon (e.g. SO9546) are good areas in the county to find the Water, 11-spot and Adonis Ladybirds, along with the inconspicuous *Coccidula scutellata*. Here, the 24-spot and 16-spot Ladybirds and *Rhyzobius litura* are also abundant.

To the north, the Wyre Forest (SO7575) has a strong population of the Scarce 7-spot Ladybird associated with nests of the Red Wood Ant, while *Clitostethus arcuatus* is locally abundant around Kidderminster (SO8378). The area is also good for the Adonis and Heather Ladybirds, while there are records of the Hieroglyphic Ladybird from Lickey Hills (SO9975) in 1991, and Hartlebury Common (SO8270), most recently in 2017.

▶ Malvern Hills, Worcestershire.

The county's network of scattered woodlands and orchards is good in general for tree-living species such as the Cream-spot and 10-spot Ladybirds, while species more restricted to coniferous trees, including the Cream-streaked, Eyed and Larch Ladybirds, have a scattered distribution but may be abundant where present.

Richard Comont

Norfolk

Norfolk has a varied mix of coastal, wetland, farmland and wooded habitats. This, and its relatively dry and sunny climate, make Norfolk a rich county for ladybirds.

The coastline can be a rewarding place to search. Species such as the Adonis Ladybird and the 11-spot Ladybird may be found in dune systems, for example at Titchwell (TF7544) and Sheringham (TG1543) on the north Norfolk coast, and at Horsey (TG4624) on the east coast.

Reedbeds in the Norfolk Broads, for example near Martham (TG4319), can provide good habitat for wetland species such as *Coccidula scutellata* and *C. rufa*.

The arable farmland of Norfolk, as elsewhere in East Anglia, can host abundant populations of the common grassland and generalist ladybirds (including 24-spot, 22-spot, 16-spot and 14-spot Ladybirds), and sweeping along the verges of fields can sometimes be fruitful.

The large areas of coniferous forest in the county (Norfolk Brecks) are very worthwhile places to look for ladybirds, harbouring many species. Thetford Forest is the largest pine woodland in lowland England,

▲ Thetford Forest, Norfolk.

occupying an area of over 19,000 hectares. The area is a plantation woodland, and much of it comprises a mix of Scots Pine and non-native Corsican Pine, with Douglas Fir and larch in lower numbers. This is a good region to look for ladybirds that specialise on pine trees, such as the Eyed Ladybird, Striped Ladybird, Cream-streaked Ladybird, 18-spot Ladybird and *Scymnus suturalis*. Many commoner species that are more generalist (including the Pine Ladybird) are also abundant. Heather Ladybirds are present, though generally less easy to locate. King's Forest (TL8274), north of the village of West Stow, is worth investigating. Sometimes a dozen or so ladybird species may be recorded here on a single day. West Stow itself has a very nice country park (TL7971) where many of the ladybirds mentioned above may be found.

Most of the forest in Norfolk is Forestry Commission owned, so access is restricted, especially where active logging is occurring. However, as with many of the UK's forested areas, Thetford Forest has a good infrastructure for walkers and cyclists and there are many paths and publicly accessible areas.

Peter Brown

Lincolnshire

Lincolnshire is a predominantly agricultural county dominated by low-lying fenland in the south, with the chalk-topped hills of the Lincolnshire Wolds running through central and eastern parts of the county. It has a long eastern coastline.

A number of notably rare ladybirds have been recorded in Lincolnshire, the majority of which are inconspicuous species. Since first being found in 2009 in the city of Lincoln (SK9568), *Clitostethus arcuatus* has subsequently been noted from the nearby village of Skellingthorpe (SK9271) and Kirkby Moor (TF2261). Two other rare ladybirds – *Scymnus femoralis* and *S. limbatus* – have been found in the north of Lincolnshire. Wolla Bank (TF5574, recorded 1997) and Little Scrubbs Wood (TF1474, recorded 1994) are the locations to search for *S. limbatus*, whereas *S. femoralis* was found in 1988 at both Scotton Beck Fields (SK8798) and Donna Nook (TF4299). A 1912 observation of *S. femoralis* in Market Rasen (TF1089) may justify a visit to the small market town on the edge of the Lincolnshire Wolds.

Of the conspicuous ladybirds resident in Britain and Ireland, a rarity in Lincolnshire is the Hieroglyphic Ladybird, recorded last in 1911. Those eager to rediscover this species in the county should search heathland at either Linwood Warren (TF1387), where it was last found, or Roxton Wood (TA1511), where the species was observed in 1909.

◀ Gibraltar Point, Lincolnshire.

Gibraltar Point National Nature Reserve (NNR) (TF5659), located on the Lincolnshire coastline approximately 5km south of Skegness, can be a rewarding site to visit in search of a range of ladybird species. The reserve comprises 445 hectares of unspoilt coastline managed by the Lincolnshire Wildlife Trust, and 16 species of ladybird have been recorded from the diverse range of habitats there. The mixed woodland can yield Eyed, Striped, Kidney-spot, Orange, Cream-streaked and 10-spot Ladybirds. Ponds and lagoons, ranging from fully freshwater to brackish conditions, provide ideal marginal habitat in which to find *Coccidula rufa*. Grassland specialists *Rhyzobius litura,* 16-spot and 22-spot Ladybirds, alongside many of the more generalist common ladybird species (such as 2-spot, 7-spot, 14-spot), can be recorded from the wildflower-rich grasslands. The generalist species may also be located on the sandy and muddy seashores and the saltmarsh areas. The Harlequin Ladybird can easily be spotted near the visitor centre and the bird hides around the reserve, while the 11-spot Ladybird can be readily found on the extensive sand dunes covered with various grasses, Elder, privet, Hawthorn and Sea-buckthorn.

Hartsholme Country Park (SK9469), located within the city of Lincoln, covers more than 80 hectares and comprises a large reservoir, Victorian landscaped gardens, woodlands and grasslands. Dry heath, *Sphagnum* bog, lakes and ponds can be found in the adjacent Swanholme Lakes Local Nature Reserve (LNR) (SK9468), a 63-hectare former gravel quarry designated a Site of Special Scientific Interest (SSSI). Twelve species of ladybird have been recorded across the two locations. The mixed woodland habitats contain over 80 different tree species, which have yielded eight of the woodland-specialist ladybird species (10-spot, Pine, Eyed, Cream-streaked, Cream-spot, Kidney-spot, Striped and Orange Ladybirds). Several of the more common, generalist species have also been recorded, but Hartsholme Country Park and Swanholme Lakes LNR are rather under-surveyed and it is highly likely that additional species may yet be found.

Alex Pickwell

South Yorkshire

South Yorkshire has a multitude of habitats, ranging from high moors, upland mires and grasslands, where the underlying rocks are overlaid with variable depths of peat, through ancient woodlands and pasture. The elevations vary from several hundred metres in the Peak District down to sea level.

Of the 47 ladybirds on the British list, 42 have been recorded in South Yorkshire, including rarities such as the Scarce 7-spot Ladybird and *Clitostethus arcuatus*. In the South Yorkshire

▶ Potteric Carr Nature Reserve, South Yorkshire.

region, the few records of the Scarce 7-spot Ladybird are invariably in association with colonies of the Northern Wood Ant.

At lower elevations, where mining subsidence has allowed flooding to occur, numerous wet areas have developed as willow carr and reedbed, for example at Potteric Carr Nature Reserve (SE5800) (a Wildlife Trust site south of Doncaster). At least 18 species of ladybird have been found there, including in 1979 the 13-spot Ladybird. The reserve, comprising some 325 hectares, is a rich mosaic of different habitats ranging from woodland and dry pasture to wetland and reedbed areas, and is home to Water Ladybirds, usually in stands of Reedmace. The Cream-spot Ladybird is regularly seen in some of the deciduous woodland areas. Another less commonly found species here is the 2-spot Ladybird, and rarer still is the Kidney-spot Ladybird.

Gillfield Wood (SK3078) is a narrow, 2km-long strip of former ancient woodland in the southwest suburbs of Sheffield that was partially clear-felled in the 1950s. Largely ignored by recorders for many years, it is now proving very interesting entomologically and is good for at least common species such as the 7-spot and Harlequin Ladybirds. Large overwintering clusters of many hundreds of Harlequin Ladybirds have been noted in houses, churches and derelict buildings in South Yorkshire, and also in dead trees and dense Ivy growths in sheltered areas. The 14-spot Ladybird is also commonly found in the region.

Derek Bateson

Cheshire

The geology of Cheshire is dominated by sandstone, with lowland plains to the west and east of the sandstone Mid Cheshire Ridge. A range of interesting wildlife habitats include forested areas, for example Delamere Forest in central Cheshire, and the edge of the Peak District in the east of the county.

Near the town of Frodsham is the almost 100-hectare Delamere Forest (SJ5471), which has a wide range of habitats including wetland, grassland and deciduous and coniferous woodland. In addition to records of various common ladybird species, there are records here of Larch, Cream-spot, Pine and Water Ladybirds (the last of these benefits from the plethora of water bodies in Cheshire and is well recorded across the county). Delamere Forest seems to be a stronghold for the Eyed Ladybird, and there are also records of some of the smaller ladybirds, such as *Scymnus auritus, S. frontalis, S. nigrinus* and *S. haemorrhoidalis.*

The Orange Ladybird is not common in Cheshire but is known from various woodland sites, such as Marbury Big Wood (SJ6576). There have been many recent Cheshire records of the Cream-streaked Ladybird.

The Hieroglyphic Ladybird has been recorded from a large number of sites in Cheshire, including Lindow Common (SJ8381), Abbotts Moss (SJ5968), Bidston Hill (SJ2889) and Wallasey sand dunes (SJ2792), with the Heather Ladybird also found at some of these sites. The Kidney-spot Ladybird is recorded from a variety of mosses and woodlands including Plumley Lime Beds (SJ7075) and Dibbinsdale LNR (SJ3482), while the Adonis Ladybird is recorded mainly from drier sites such as Wallasey sand dunes, although it has also been recorded well inland.

Of the inconspicuous ladybirds, *Coccidula rufa* is very well distributed in the county, whereas there is a paucity of records for *C. scutellata*, most records for which come from the Delamere area. Likewise, *Rhyzobius litura* is very common, whereas *R. chrysomeloides* is seldom recorded and *R. lophanthae* has not been recorded in the county. A schools ladybird project at the RECORD Local Biological Records Centre increased recording and led to the addition of several recent ladybird records. These included *Scymnus femoralis* and *S. auritus* each at two locations, *S. frontalis*, and many new sites for *Nephus redtenbacheri.*

Don Stenhouse

48

▶ Delamere
Forest,
Cheshire.

Lancashire

The Forest of Bowland dominates much of the north of Lancashire. A band of plains and valleys separates this from the Southern Pennines, parts of which occupy the southeast of the county. A substantial western coastline, including the large estuary of Morecambe Bay, has important habitats for wildlife.

The extensive dune systems of the Sefton coast have produced many ladybird records over the years, including the scarce species *Scymnus femoralis*, which was recorded from Freshfield (SD2908) many years ago. There are no other apparent records of that species for Lancashire, so it may be interesting to see if it can still be found at Freshfield, for example at the base of grasses. Other small ladybirds recorded in the area are *S. auritus* (also recorded many years ago in Silverdale, SD4575), and *S. frontalis* and *S. suturalis* (also recorded at St Anne's, SD3228, and other locations in the area). The Eyed Ladybird can be found at locations within the Forest of Bowland, such as Marshaw (SD5953). There are extensive pine woodlands along the Sefton coast, and the Pine Ladybird has been recorded here, along with the Larch Ladybird.

The Scarce 7-spot Ladybird has been recorded from the adjacent county of Cumbria, and could possibly be found just within the Lancashire boundary at Gait Barrows NNR (SD4777), which has numerous Red Wood Ant colonies. Another rarity, the 5-spot Ladybird, has been

▲ Ainsdale on Sea, Sefton Coast, Lancashire.

recorded in south and north Lancashire but there appear to be few records, and none are recent, although the numerous tributaries of the Ribble and other rivers may provide suitable habitat.

The Hieroglyphic Ladybird is quite scarce in the county and has only been recorded from the few mosses, including Chat Moss and Astley Moss (SJ7096) and Holcroft Moss (SJ6893). Also scarce is the Cream-streaked Ladybird, which seems to be confined mainly to the Sefton coast at Formby Point (SD2706) and other nearby areas with pine woodland. Also recorded mainly on the coast is the Adonis Ladybird, while the closely related 13-spot Ladybird has been recorded in Lancashire in the Cleveleys and Preston areas but is genuinely scarce.

The Heather Ladybird is very sparsely recorded in Lancashire and seems to be confined to the Freshfield area, possibly as a result of drainage and peat extraction of mosslands, while the Kidney-spot Ladybird is much more widely recorded. The 11-spot Ladybird is well recorded in south and mid-Lancashire, with lots of records from coastal sites such as Oglet shore (SJ4481) and the more inland Gait Barrows (SD4877). *Coccidula scutellata* is barely known in the county, with just a few records.

Don Stenhouse

Wales

Meandering rivers, expansive mountains and small pockets of ancient woodland ensure there is an abundance of suitable habitat for ladybirds in Wales – from the mountain ranges of Snowdonia, the Brecon Beacons and the Cambrians to the island of Anglesey and the valleys of South Wales, Elan Valley in mid-Wales and the Wye Valley in the east. Several common species such as the Orange, 10-spot, 2-spot, 7-spot and 14-spot Ladybirds are widespread. Although more commonly recorded in England, in Wales the 11-spot Ladybird tends to prefer coastal habitats, for example in Pembrokeshire at Freshwater West Beach (SR8899) and Tenby (SN1203), and in Anglesey at Cemlyn Bay (SH3392) and Newborough (SH4265). This species can also sometimes be found further inland, near rivers.

▲ Near Glasbury, Powys.

Several inconspicuous ladybirds occur on Anglesey. *Scymnus limbatus* can be found on willow and elm trees, while it is worth searching low vegetation for *S. schmidti*. *Hyperaspis pseudopustulata* may prove more difficult to find, but low vegetation in coastal areas of Anglesey tends to be the best place to locate this species.

For more of a challenge, it may be worth searching the southwest Pembrokeshire coast for the Heather and Water Ladybirds, each of which has been recorded only sporadically in Wales in the last decade. Also with a coastal preference is *Rhyzobius litura*, a common species on low vegetation but one that can be elusive because of its cryptic colouration.

Glasbury (SO1739) in the Wye Valley is another place where a ladybird enthusiast may come across species not often recorded in Wales. Here, when searching grassland, the 16-spot Ladybird may be found. The Adonis Ladybird has not often been recorded in Wales, but searching at Glasbury along shingle banks may yield success, specifically on thistles, Cow Parsley or Nettle.

Wales is a stronghold for one of Britain's rarest ladybirds, the 5-spot Ladybird, inhabiting unstable river shingle banks. It prefers sparse vegetation such as thistles and Nettle on the shingle banks, and on hot days it may also be seen scurrying across the shingle itself. The 5-spot Ladybird can be found on the banks of the Rivers Wye, Usk and Severn and the Afon Twyi, Ystwyth and Rheidol; specific areas include Glasbury (SO1739) or Hay-on-Wye (SO2242) in the Wye Valley and Abergavenny (SO2913) on the River Usk.

Rachel Farrow

Central Scotland

Thirty-one species of ladybird have been recorded in Scotland, of which only 25 have been recorded since 1997. Of these 25 species, 22 are conspicuous and just 3 are inconspicuous species. The recording intensity in Scotland is fairly low in comparison to some other areas of Britain, so it is likely that other species may be present.

▲ Fannyside Moss, Lanarkshire.

Central Scotland is formed of a rift valley between the Scottish Highlands and the Southern Uplands and generally contains low-lying, flat land. Known as the Central Belt, this area has the highest human population density in Scotland. As a result, it contains ladybird species common in urban areas, for example 2-spot and 10-spot Ladybirds. The 14-spot, Cream-spot and 7-spot Ladybirds are commonly found across the Central Belt, although they are recorded less often than south of the border. The 22-spot Ladybird is, however, rarely found in Scotland.

The Harlequin Ladybird was first recorded in Scotland in 2007 and has since been reported as far north as Orkney. However, the climatic conditions associated with much of Scotland (high precipitation and low temperatures) appear to restrict reproduction of the Harlequin to urban areas, and while some records of breeding have been reported from Edinburgh, these are rare.

As well as urban habitats, central Scotland has many other habitat types, including grassland, moorland, peatland and wetlands. In North Lanarkshire, Fannyside Moss (NS7973) is an area of peatland within the Slamannan Plateau that is both an SSSI and a Special Area of Conservation (SAC). Although it has been subject to peat extraction in the past, recent restoration has been undertaken to maintain the natural bogs and biodiversity at this site. Of particular interest are the records of the heathland specialist Hieroglyphic Ladybird. While this species has been recorded across the Central Belt, 64 of the 187 Scottish reported sightings are from Fannyside Moss, so this restored area is clearly a promising habitat for the species.

To the east, Edinburgh is bordered by West Lothian, Midlothian and East Lothian. Much of the land in this area is agricultural, with coastal habitats also present. Grassland species such as the 14-spot and 7-spot Ladybirds and *Nephus redtenbacheri* have been reported here. Despite the abundance of coastal (including dune and Gorse), coniferous and heathland habitats, species adapted to these habitats – such as the Cream-streaked and 24-spot Ladybirds – have not been recorded. In East Lothian, the Orange Ladybird is commonly found associated with Sycamore trees around the coastal forests.

The River Forth leading to the Firth of Forth on the east of the Central Belt is designated as an SSSI. This coastal habitat is important for biodiversity, particularly shorebirds, and is also a good habitat for the 11-spot Ladybird. In particular, this species has been recorded occurring upriver in the marshy habitats of Tullibody Inch (NS8692) and Alloa Inch (NS8791), as well as in a lowland heath habitat near Fallin (NS8391).

In addition to common generalist species, ladybirds not normally associated with urban environments have been found in the Glasgow area, including the Larch and Striped Ladybirds, both normally associated with coniferous woodland. Although very rare in Scotland, sightings of the Water Ladybird have been reported from the Caerlaverock NNR near Dumfries (NY0365), which contains the wetland and reedbed habitat favoured by this species.

Katie Berry

Ireland

The west of Ireland has a rugged coastline and is quite mountainous. Much of central Ireland comprises low-lying plains. There are large lakes and wetland areas, and important woodlands, though Ireland has a slightly lower proportion of forested land than does Britain.

Ballybrack Woods (W6968) (known locally as the 'Mangala') is located in the Douglas suburb of Cork city. The woods have a 1.2km-looped footpath/cycle path, which is listed as an Irish Trail. Ballybrack Woods consists of a semi-urban mixed deciduous ribbon woodland, located within a V-shaped valley and the Ballybrack River, and is a good place to look for wildlife, with the valley floor especially good for 22-spot, 7-spot and 14-spot Ladybirds. The riparian edges are also good places to find 10-spot and Orange Ladybirds on the underside of the leaves of trees such as Sycamore.

Doneraile Wildlife Park (R6007) in the northern part of County Cork comprises 166 hectares of parkland, grazed pasture, walkways and the River Awbeg. Accessible on foot, the parkland holds an extensive range of mature and specimen trees such as limes, oaks, Sycamore, Yew and Plane. Orange Ladybirds may be seen on the underside of Sycamore leaves here, as well as 14-spot and 10-spot Ladybirds on the oaks along the eastern side of the park.

In west Cork, Glengarriff Woods Nature Reserve (V9157) covers some 300 hectares. The woods form one of the best examples of oceanic Sessile Oak woodland in Ireland and the site is part of a wider SAC. Glengarriff Woods is managed primarily for conservation and amenity purposes, and along the paths ladybirds such as the 7-spot and 14-spot can be found in the open areas, in the trees and along path edges.

Fota Wildlife Park (W7870) on Fota Island to the east of Cork city covers an area of 40 hectares and is managed by the Zoological Society of Ireland. A series of paths runs around the park, and several ladybird species including Orange, 10-spot, 7-spot, 14-spot and Kidney-spot can be found on isolated trees such as oaks and Sycamore. In addition to these native ladybird species, the Harlequin Ladybird can be seen throughout the park, especially during August and September.

Gill Weyman

▶ Ballybrack Woods, Douglas, Cork.

Using this field guide

The *At-a-glance guide* to the species of ladybirds (pages 62–67) should provide a useful starting point for identification. However, both the conspicuous and inconspicuous ladybirds can be tricky to identify in some cases.

Conspicuous ladybirds

It is often assumed that the brightly coloured and strongly patterned ladybirds will be easy to identify. This is certainly the case for some species, but not for all. Indeed, there are a number of species that are relatively easily confused with one another (see *Confusion species* within the species accounts and the *Similar species* section, pages 146–151). It is generally not a single characteristic that separates one species from another, but more often a combination of features. For the conspicuous ladybirds, noting the size can be helpful, but also the patterning on the top of the thorax (pronotum) can be more instructive than the elytral patterning.

It can also be extremely useful for identification purposes (and indeed further study) to note the habitat in which the ladybird is found. Some species generally occur in specific habitats such as coniferous woodlands or heathlands. Of course, there will always be exceptions, but even so, location can provide a starting point for identification.

▲ A typical conspicuous ladybird species: 7-spot Ladybird.

During early summer adult ladybirds become scarce, while the immature stages dominate. As described previously, the larval stage consists of four instars, the stages between each cycle of moulting. We have provided information on late-instar larvae and pupae in the species accounts, but it is very difficult, and in many cases almost impossible, to identify early-instar larvae. For the late-instar larvae it can be instructive to look at colour and any patterning, but also to examine the way in which the bristles or spines are organised. These bristles and spines are often still apparent on the larval skin that remains attached to the base of the pupa, so they can also be useful in this context. Although the pupae often have distinct colouration and sculpturing, they can be difficult to identify to species. It should be noted that the amount of melanin (black pigment) decreases with developmental temperature, so some pupae of a particular species are darker than others.

55

Inconspicuous ladybirds

The inconspicuous ladybirds are, as their name suggests, much smaller and less obviously ladybird-like than the conspicuous ladybirds. However, many can be identified to species in the field with a good hand lens. Perhaps the greatest challenge is seeing them in the first place. It is always worth looking carefully through the debris in a sweep net or beating tray (see *Equipment*, pages 29–32) for any small moving creatures that could be inconspicuous ladybirds, often superficially dull in appearance because of their hairs. While some are genuinely uncommon, there is no doubt that most species of inconspicuous ladybird are under-recorded, and we hope that this field guide will encourage recording of these tiny beetles. Many of the larvae of inconspicuous ladybirds are very similar to one another, and so only the adult stages are illustrated in this field guide.

If you are unfamiliar with inconspicuous ladybirds then it is worth having your identification checked by another recorder. There are a number of species that are expanding in range or are recent arrivals, and this adds to the complications of identification. The key within Roy *et al.* (2013) is a useful reference for those wishing to take identification of inconspicuous species further.

▶ A typical inconspicuous ladybird species: *Rhyzobius litura*.

Species accounts

The taxonomic order of British Coccinellidae follows the checklist compiled by Duff (2018), with 47 species of ladybird included in this guide. There have been a number of new arrivals in recent decades, and some further species are predicted to arrive in due course (see *Potential new species*, pages 152–153).

SCIENTIFIC NAMES AND COMMON NAMES
We have listed the species within tribes (but starting with the conspicuous ladybirds) and then in order according to Duff (2018). Scientific names for even well-studied ladybirds can change as new evolutionary relationships are revealed. This can be viewed either as frustrating or as exciting! It certainly highlights the outcomes of detailed and meticulous studies by dedicated people attempting to unravel the complexities of relationships between species in order to derive an accurate phylogenetic tree. The current accepted name is given in the heading of each species account.

   We have included common (vernacular) names for all of the species. While common names have been in use for the conspicuous ladybirds for several decades, the inconspicuous ladybirds have not previously been given common names. There can be considerable disagreement as to whether or not such names should be used and how they should be decided upon. However, we have taken the step of discussing potential names with a number of beetle enthusiasts and agreed on the names given in the species account headings. We have tried to keep these simple and descriptive, but in some cases we have used the name of the genus within the common name, to provide some link and familiarisation with the scientific name.

ILLUSTRATIONS
The *At-a-glance guide* (pages 62–67) shows the most common adult colour form of each species, arranged in taxonomic order (Duff 2018) with the size of the species in proportion to one another (although the magnification differs between the conspicuous and inconspicuous species). The species accounts include illustrations of late-instar larvae, pupae and adults for the conspicuous ladybirds, and adults only for the inconspicuous ladybirds because of the difficulty of finding the immature stages of these small insects. We have included examples of the most common colour forms, but it is important to note the extreme variability in some species; even variation in spot size can make individuals with the same colour form appear quite different. A life-size image has been provided to give an indication of scale. The species are depicted from the top view. Colour photographs have been provided for the conspicuous species, but these are not readily available for all the inconspicuous ladybirds.

DISTRIBUTION MAPS
The species distribution maps have been produced using records received by the UK Ladybird Survey (formerly known as the Coccinellidae Recording Scheme) from 1975 up to the end of 2015. An approach has been used that estimates the distribution in a systematic and repeatable way by filling gaps in the recording by joining data points, based on their distance to neighbouring points. Ladybirds, like many insects, are on the move – and so the coming years are likely to show expansion northwards and within Ireland.

   The data used up to the end of 2010 represent a compilation of records from many sources as described in Roy *et al.* (2011). The main distinction between the dataset published in Roy *et al.* (2011) and updates from 2011 to 2015 is that in the latter period most of the data are directly from the UK Ladybird Survey, whereas previously a much greater range of data sources was included. There is certainly merit in combining datasets from all sources going forward, but the number of records received through the UK Ladybird Survey is extremely high and provides sufficient coverage to generate the stylised distribution maps. The summary maps (10km resolution) opposite provide an overview of the overall coverage of records, number of records and number of species from the data compilation described above.

No. Records
- 1–2
- 3–10
- 11–30
- 31–99
- 100+

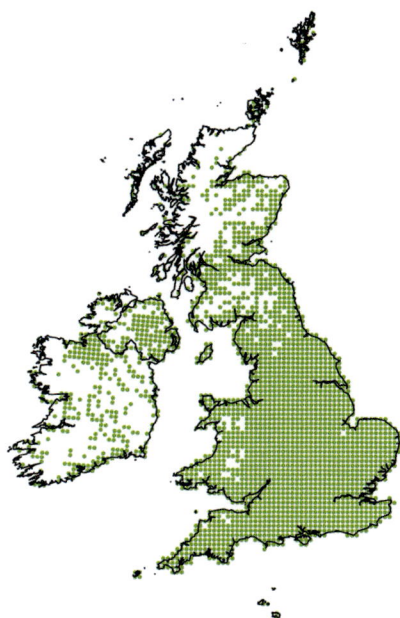

▲ Overall coverage of ladybird records (1975–2015).

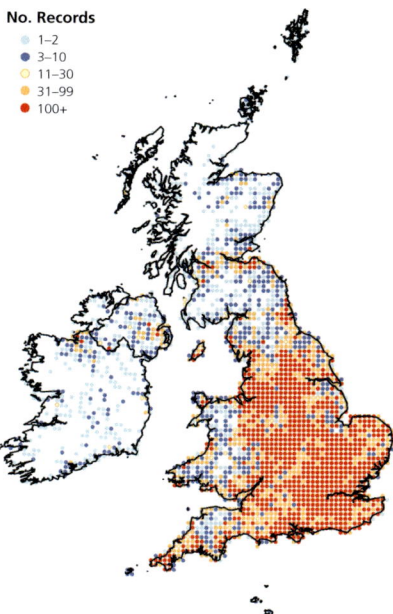

▲ Intensity of recording (number of records per 10km square).

GENERAL DESCRIPTION

The general description of each species includes key features for identifying larvae, pupae and adults of the conspicuous ladybirds, and adults of the inconspicuous ladybirds. The characters should be readily observable on individuals found in the field, but some require the use of a hand lens, especially on small species.

The key features for the adults are **length**, elytral **background colour**, **pattern**, **number of spots**, and **spot fusions** (conspicuous species only); the colouration of the **pronotum** and **legs** can also be diagnostic. Where relevant, **other features** are also described. Some species are variable in colour pattern, and these polymorphisms are described both in terms of elytral colours (**other colour forms**) and also **spot fusions** (conspicuous species only). The spot number can also vary, so the range is given, with the most common spot number in parentheses.

The descriptions of the larvae relate to third and fourth instars only. For both larvae and pupae, the descriptions provide the major distinguishing features.

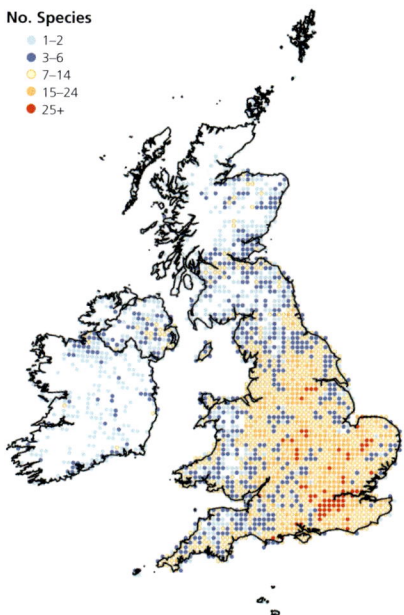

No. Species
- 1–2
- 3–6
- 7–14
- 15–24
- 25+

▲ Species richness (number of species recorded per 10km square).

CONFUSION SPECIES

Within the species accounts, we have provided tables of characteristics for separating species that are easily confused with one another. For example, the Kidney-spot Ladybird is sometimes confused with the *conspicua* colour form of the Harlequin Ladybird. While the Kidney-spot Ladybird is smaller with black legs, perhaps the most striking difference can be seen by comparing the pronotum of each species: the Harlequin Ladybird (in all colour forms) has white markings at least on either side of the pronotum and sometimes centrally too, whereas the Kidney-spot Ladybird has an entirely black pronotum.

FOOD

General information is provided on feeding behaviour within the Introduction (see *Feeding*, pages 20–24), but for each species we document the most common food types within the species accounts (noting that larvae and adults have similar food preferences). Most species of ladybird will consume a variety of food and it would be fascinating to capture the diet breadth of the predatory ladybirds from field studies over time. This could be particularly revealing as availability of different prey is likely to vary with climate change. The organisers of the UK Ladybird Survey would be very pleased to collate new information on feeding behaviour, and observations on the interactions between ladybirds and their food would be welcomed (see *Websites, apps and social media*, page 154).

HABITAT

The habitat in which a ladybird is found (see *Ladybird habitats*, pages 34–39) can provide useful information for identification, and so the common habitats and details of resources (such as specific plants) within those habitats are provided for each species. This information is given for both the active periods of the ladybird's life cycle and also for overwintering.

SUGGESTED SURVEY METHOD

For many ladybirds a range of survey methods is appropriate, but we have suggested those that we have found most effective for sampling each species. These align with the descriptions provided in the section on *Equipment* (pages 29–32).

RANGE

The current distribution is shown on the distribution maps, but we have also provided a description of the extent of the species within Britain and Ireland. In addition, we have suggested likely future changes in the distribution as a consequence of climate change.

NATIONAL STATUS

Many different organisations have produced a plethora of lists of species conservation status over the last 30 years, including Red Lists, Biodiversity Action Plan Priority Lists, species listed on European Directives, species listed on the Schedules of the Wildlife and Countryside Act (1981), and various lists of rare and scarce species. The Joint Nature Conservation Committee (JNCC) has collated many of the current lists into a downloadable spreadsheet of species designations (http://jncc.defra.gov.uk/page-3408). Twelve species of ladybird appear in this spreadsheet because they have been assigned a conservation designation; these are shown in the table opposite. This is in need of a review, in conjunction with the new trends data that we show here. There are currently no ladybirds on the lists of Species of Principal Importance (which replace the UK Biodiversity Action Plan Priority Lists).

In Roy *et al.* (2011) a system was devised, based on previous studies from other recording schemes including the Bees, Wasps and Ants Recording Scheme (BWARS), to include the 35 species of ladybird that do not have a conservation designation and provide an indication of status. In brief, there are 3,870 10km squares across Britain, Ireland and the Channel Islands. For each species the number of 10km squares in which a species was recorded was converted to a percentage of this total, and the species were subsequently categorised as follows:

Ubiquitous – found almost everywhere and in at least 25% of the 10km squares;
Very widespread – found in fewer than 25% but more than 20% of the 10km squares;
Widespread – fewer than 20% but more than 10% of the 10km squares;
Local – fewer than 10% but more than 5% of the 10km squares;
Very local – fewer than 5% of the 10km squares.

In this book we apply this to all 47 species, but it is important to note that some of the species, particularly the inconspicuous ladybirds, are under-recorded, and therefore the estimate of status may be inaccurate (see table overleaf).

Conservation designations for the 12 species of ladybird included in the JNCC Conservation Designations Spreadsheet (JNCC 2018)

Species	Reporting category	Conservation designation
Clitostethus arcuatus	Red Listing based on pre-1994 IUCN guidelines	Endangered
Scymnus femoralis	Rare and scarce species (not based on IUCN criteria)	Nationally scarce/Nb
Scymnus schmidti	Rare and scarce species (not based on IUCN criteria)	Nationally scarce/Nb
Scymnus limbatus	Rare and scarce species (not based on IUCN criteria)	Nationally scarce/Nb
Nephus quadrimaculatus	Red Listing based on pre-1994 IUCN guidelines	Vulnerable
Nephus bisignatus	Red Listing based on pre-1994 IUCN guidelines	Extinct
Hyperaspis pseudopustulata	Rare and scarce species (not based on IUCN criteria)	Nationally scarce/Nb
Platynaspis luteorubra	Rare and scarce species (not based on IUCN criteria)	Nationally scarce/Na
13-spot Ladybird *Hippodamia tredecimpunctata*	Red Listing based on pre-1994 IUCN guidelines	Insufficiently known
Adonis Ladybird *Hippodamia variegata*	Rare and scarce species (not based on IUCN criteria)	Nationally scarce/Nb
Scarce 7-spot Ladybird *Coccinella magnifica*	Rare and scarce species (not based on IUCN criteria)	Nationally scarce/Na
5-spot Ladybird *Coccinella quinquepunctata*	Red Listing based on pre-1994 IUCN guidelines	Rare

Endangered – occurring only as a single population or otherwise in danger of extinction;
Vulnerable – declining or in vulnerable habitat and likely to become endangered in the near future;
Rare – very restricted by area or by habitat or with thinly scattered populations, occurring in no more than 15 10km squares;
Nationally scarce – uncommon in Britain, with two grades of rarity: Notable a (Na) – occurring in 16–30 10km squares; Notable b (Nb) – occurring in 31–100 10km squares.
IUCN – International Union for Conservation of Nature.

DISTRIBUTION TRENDS

Ecologists are now using advanced statistical modelling techniques to look at trends in ladybird distributions over time. The records compiled through the UK Ladybird Survey and other sources essentially represent repeat visits to 1km squares, and so from this we can estimate species distribution, while accounting for differences in species detectability. This is essential because biological recording can be quite *ad hoc* and this can cause biases in trend estimates.

Number of 10km squares and records within the BRC Coccinellidae (ladybird) database for each species (ordered by number of 10km squares) from 1975 to 2015

Species	Number of 10km squares	Number of records	Status	Distribution trend
7-spot Ladybird	2058	37129	Ubiquitous	Decreasing
Harlequin Ladybird	1383	34235	Ubiquitous	Increasing
14-spot Ladybird	1374	12127	Ubiquitous	Stable
10-spot Ladybird	1328	6760	Ubiquitous	Stable
2-spot Ladybird	1299	16029	Ubiquitous	Decreasing
Cream-spot Ladybird	1036	4456	Ubiquitous	Stable
Orange Ladybird	959	5053	Very widespread	Increasing
22-spot Ladybird	927	5389	Very widespread	Decreasing
Coccidula rufa	839	2704	Very widespread	Decreasing
Rhyzobius litura	763	2621	Widespread	Stable
11-spot Ladybird	756	1951	Widespread	Decreasing
Kidney-spot Ladybird	710	2270	Widespread	Stable
Pine Ladybird	695	5520	Widespread	Stable
24-spot Ladybird	637	3442	Widespread	Stable
16-spot Ladybird	569	2865	Widespread	Stable
Larch Ladybird	537	1146	Widespread	Decreasing
Eyed Ladybird	518	1261	Widespread	Decreasing
Water Ladybird	466	1829	Widespread	Decreasing
Adonis Ladybird	371	1151	Local	Stable
Cream-streaked Ladybird	301	948	Local	Stable
18-spot Ladybird	218	527	Local	Stable
Heather Ladybird	208	510	Local	Stable
Hieroglyphic Ladybird	186	395	Very local	Decreasing
Striped Ladybird	184	325	Very local	Stable
Scymnus suturalis	179	384	Very local	Stable
Scymnus frontalis	163	282	Very local	Stable
Nephus redtenbacheri	162	227	Very local	Decreasing
Scymnus auritus	132	199	Very local	Decreasing
Coccidula scutellata	130	232	Very local	Decreasing
Scymnus haemorrhoidalis	103	145	Very local	Decreasing
Rhyzobius chrysomeloides	65	168	Very local	Increasing
Stethorus pusillus	64	84	Very local	Decreasing
Nephus quadrimaculatus	55	144	Very local	Stable
Scymnus schmidti	48	80	Very local	
Scarce 7-spot Ladybird	41	130	Very local	Stable
Hyperaspis pseudopustulata	40	61	Very local	
5-spot Ladybird	39	156	Very local	Stable
Platynaspis luteorubra	37	77	Very local	
Scymnus femoralis	36	48	Very local	
Scymnus limbatus	26	35	Very local	
Scymnus interruptus	26	114	Very local	Stable
Rhyzobius lophanthae	25	49	Very local	
13-spot Ladybird	20	35	Very local	

Species	Number of 10km squares	Number of records	Status	Distribution trend
Scymnus nigrinus	17	25	Very local	
Bryony Ladybird	17	162	Very local	Stable
Clitostethus arcuatus	16	31	Very local	
Nephus bisignatus	1	1	Very local	

Those species for which no distribution trend is given have insufficient records for robust analysis. In total there are 3,870 10km squares.

A statistical technique called 'occupancy modelling' can be used to estimate the trends in species distribution over time.

The trends presented in each of the species accounts in this field guide were produced by Charlie Outhwaite and Nick Isaac (Centre for Ecology & Hydrology). For this analysis, records from 1km grid cells with a specific date were used. This enabled assessment of fine-scale change, while using as many of the records as possible within the analysis. By rearranging these records, according to their unique combination of specific dates and location (1km squares), an occupancy modelling framework can be used to estimate the proportion of sites within the UK occupied by a species for each year. From this, distribution trends can be determined for each species by looking at the difference between the first and last year of a period of interest.

In the species accounts, the trends are presented as **stable**, **increasing** or **decreasing** for the period 1995–2015. It is important to note that these are distribution trends rather than population trends, the latter requiring abundance to be measured. Furthermore, these trends are for the UK only (not Ireland or the Channel Islands), but there would be merit in extending these across the study area represented by this field guide. It would also be interesting to assess the trends at smaller spatial scales in order to give regional or county trends. Further work is required to decipher the factors that are influencing these trends, but from previous correlative studies it is known that climate, land use and the arrival of the Harlequin Ladybird are all playing a part.

▲ A Harlequin Ladybird (top) in the same habitat as two forms of Cream-streaked Ladybird.

At-a-glance guide

TRIBE CHILOCORINI

line of red
spots

common on
heathlands

Heather Ladybird
Chilocorus bipustulatus
p.68

two large
red spots

Kidney-spot Ladybird
Chilocorus renipustulatus
p.70

front red spots
comma-shaped

Pine Ladybird
Exochomus quadripustulatus
p.72

pronotum
with distinct
cream margin

association with
ants

Ant-nest Ladybird
Platynaspis luteorubra
p.74

TRIBE COCCINELLINI

white spots in
rows running
lengthways
down elytra

deciduous trees

Orange Ladybird
Halyzia sedecimguttata
p.76

low vegetation,
grasslands

22-spot Ladybird
Psyllobora vigintiduopunctata
p.78

elongate

reedbeds and
grassland in
marshy habitats

Water Ladybird
Anisosticta novemdecimpunctata
p.80

side row of
spots often
fused in a line

grassland

16-spot Ladybird
Tytthaspis sedecimpunctata
p.82

distinctive pronotum marking with black spot on either side

elongate

marshy lowland habitats

very rare

13-spot Ladybird
Hippodamia tredecimpunctata
p.84

×6

spots often towards rear of elytra

often on bare ground or gravel

Adonis Ladybird
Hippodamia variegata
p.86

63

light brown and slightly speckled

conifer specialist – mainly larch

Larch Ladybird
Aphidecta obliterata
p.88

variable species but often two large black spots

2-spot Ladybird
Adalia bipunctata
p.90

one spot on each 'shoulder'

10-spot Ladybird
Adalia decempunctata
p.93

variable species with black stripes, spots and patches

Heather heathland

Hieroglyphic Ladybird
Coccinella hieroglyphica
p.96

central spots often large

four small white markings on underside by legs

associated with wood ant nests

Scarce 7-spot Ladybird
Coccinella magnifica
p.98

×6

disturbed
river shingle

5-spot Ladybird
*Coccinella
quinquepunctata*
p.100

two small
white
markings on
underside
by legs

7-spot Ladybird
Coccinella septempunctata
p.102

often
associated
with coastal
habitats

11-spot Ladybird
*Coccinella
undecimpunctata*
p.104

two spots on
each 'shoulder'

Harlequin Ladybird
Harmonia axyridis
p.106

speckled pronotum
pattern

distinct spot
pattern of either
16 or (less
commonly) 4 spots

cream streaking on
wing-cases

conifer specialist

Cream-streaked Ladybird
Harmonia quadripunctata
p.109

rectangular
spots

14-spot Ladybird
Propylea quattuordecimpunctata
p.112

×6

eye-like
markings on
pronotum

often pale
rings around
spots

conifer specialist

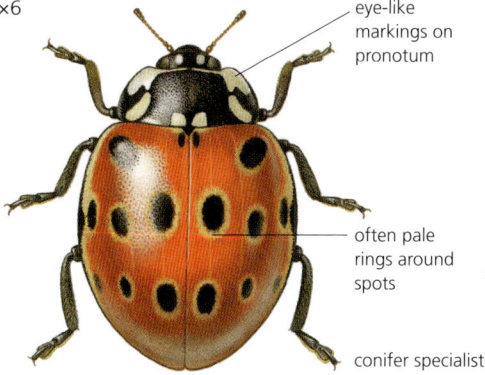

Eyed Ladybird
Anatis ocellata
p.114

distinctive
cream
marking at
join of elytra
behind
pronotum

conifer
specialist

18-spot Ladybird
Myrrha octodecimguttata
p.116

cream spots in
rows running
transversely
across elytra

deciduous trees

Cream-spot Ladybird
*Calvia
quattuordecimguttata*
p.118

cream stripes
and spots

conifer
specialist

Striped Ladybird
Myzia oblongoguttata
p.120

TRIBE EPILACHNINI

slightly hairy

gardens with
White Bryony

Bryony Ladybird
Henosepilachna argus
p.122

slightly hairy

rough grass

24-spot Ladybird
*Subcoccinella
vigintiquattuorpunctata*
p.124

×8

TRIBE COCCIDULINI

elongate and rather
flattened

marshes, riversides,
pondsides

66

Red Marsh Ladybird
Coccidula rufa
p.126

distinct triangular
marking behind
pronotum

marshes, riversides,
pondsides

Spotted Marsh Ladybird
Coccidula scutellata
p.127

U-shaped mark
towards end of
elytra

coniferous and
deciduous trees

Round-keeled Rhyzobius
Rhyzobius chrysomeloides
p.128

U-shaped mark
towards end of
elytra

grassland;
low-growing
vegetation

Pointed-keeled Rhyzobius
Rhyzobius litura
p.129

dull orange
head and
pronotum

coniferous and
deciduous trees

Red-headed Rhyzobius
Rhyzobius lophanthae
p.130

TRIBE SCYMNINI

yellow-cream
horseshoe mark

coniferous and
deciduous trees

Horseshoe Ladybird
Clitostethus arcuatus
p.132

orange-red
spots at tip of
elytra

often coastal or
wet habitats

False-spotted Ladybird
Hyperaspis pseudopustulata
p.131

reddish-brown
spots towards
tip of elytra

probably extinct
in Britain

Two-spotted Nephus
Nephus bisignatus
p.133

reddish
kidney-shaped
spots; anterior
larger than
posterior pair

often where
Ivy is prevalent

Four-spotted Nephus
Nephus quadrimaculatus
p.134

×8

large irregular red
patches on sides
of elytra

often coastal

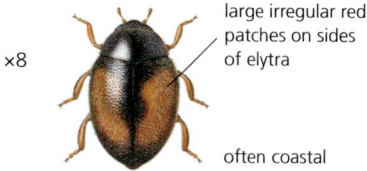

Red-patched Nephus
Nephus redtenbacheri
p.135

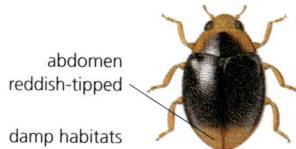

abdomen
reddish-tipped

damp habitats

Red-rumped Scymnus
*Scymnus
haemorrhoidalis*
p.136

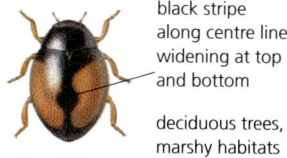

black stripe
along centre line
widening at top
and bottom

deciduous trees,
marshy habitats

Bordered Scymnus
Scymnus limbatus
p.137

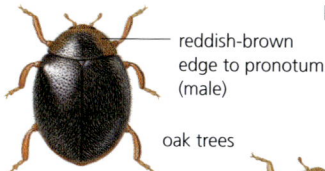

reddish-brown
edge to pronotum
(male)

oak trees

Oak Scymnus
Scymnus auritus
p.138

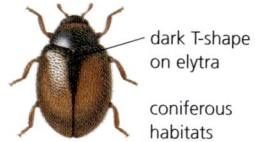

dark T-shape
on elytra

coniferous
habitats

Pine Scymnus
Scymnus suturalis
p.139

reddish-brown legs,
antennae
and mouthparts

heathlands and
dry habitats

Heath Scymnus
Scymnus femoralis
p.140

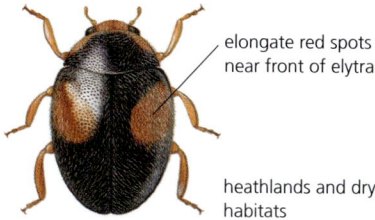

elongate red spots
near front of elytra

heathlands and dry
habitats

Angle-spotted Scymnus
Scymnus frontalis
p.141

triangular red spots
to edge of elytra

Red-flanked Scymnus
Scymnus interruptus
p.142

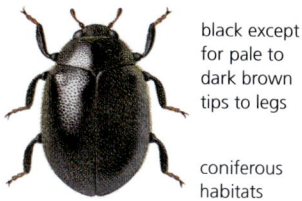

black except
for pale to
dark brown
tips to legs

coniferous
habitats

Black Scymnus
Scymnus nigrinus
p.143

mostly
coastal

Schmidt's Scymnus
Scymnus schmidti
p.144

yellowish
legs

Dot Ladybird
Stethorus pusillus
p.145

Heather Ladybird

Chilocorus bipustulatus
(Linnaeus, 1758)

Small black ladybird with a transverse line of red spots across the elytra. The species name *bipustulatus* suggests two red spots (pustules) but these are fragmented and so often appear as a line of six red spots. The rim around the edge of the elytra is a distinctive feature of some species within this tribe, giving the appearance of a small bowler hat.

Identification
Adult
Length 3–4mm
Background colour Black
Pattern Red spots
Number of spots 2–6 (6)
Spot fusions Common
Other colour forms None
Pronotum Black
Leg colour Dark brown/black
Other features Distinct rim around the edge of the elytra

Late-instar larva Dark brown, with tubercles bearing long black spines giving rise to hairs with extensive side-branching. The pale band (first abdominal segment) appearing midway along the larva is characteristic.

Pupa Larval skin encloses the lower part of the pupa, and the pale first abdominal segment is often apparent and diagnostic.

Food
The Heather Ladybird is a predatory ladybird that feeds on scale insects. Ladybirds of the genus *Chilocorus* have unique adaptations (including modified mandibles) that enable them to feed on armoured scale insects (diaspidids) that live under a waxy scale for much of their life cycle.

Habitat
The Heather Ladybird is commonly found on heathland, but there are also a number of coastal records from dune systems and scrub, where it can be found on shrubs such as Bracken, Bramble and Gorse and also on trees including sallow, willow and Scots Pine. An increasing number of reports of this species have been associated with Leyland Cypress, where scale insects can be abundant.

▲ Heather Ladybird.

life size

adult

larva

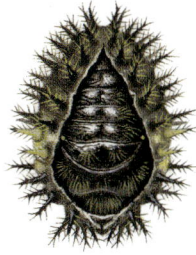

pupa

Suggested survey method
Sweep netting in heathland habitats or visual searching on sunny days.

Range
Historically this species has been limited to England with scattered records in Wales and across Ireland. However, recent records from Scotland suggest it is expanding northwards.

National conservation status
Local.

Distribution trend (1995–2015)
Stable.

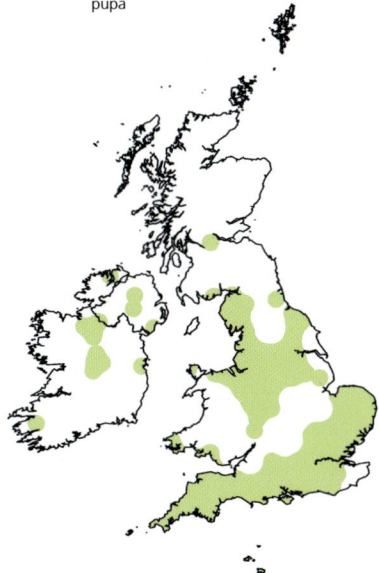

Kidney-spot Ladybird   *Chilocorus renipustulatus*
(Scriba, 1791)

Distinctively domed medium black ladybird with two large red spots and rim around edge of elytra. Often seen on the bark of Ash trees.

Identification
Adult
Length 4–5mm
Background colour Black
Pattern Red spots
Number of spots 2
Spot fusions None
Other colour forms None
Pronotum Black
Leg colour Dark brown/black
Other features Distinct rim around the edge of the elytra

Late-instar larva Pale to dark reddish-brown with distinctive long black bristles giving rise to extensive side-branches and overall spiky appearance of larva.

Pupa Black and shiny but with characteristic larval skin enclosing the lower part of the pupa.

Confusion species
The Kidney-spot Ladybird is commonly confused with the Harlequin Ladybird f. *conspicua*. However, the Kidney-spot is smaller than the Harlequin and the pronotum is entirely black, whereas the pronotum of the Harlequin is white with black markings.

Food
The Kidney-spot Ladybird is a predatory ladybird that feeds on scale insects.

Confusion species		
	Kidney-spot Ladybird	Harlequin Ladybird f. *conspicua*
Size	4–5mm	5–8mm
Pronotum	Black	White with black M-mark or solid trapezoid marking
Leg colour	Dark brown/black	Orange to brown

▲ Kidney-spot Ladybird.

70

adult

life size

larva

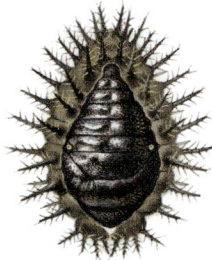
pupa

Habitat

This species is commonly found in deciduous woodland, but also in mixed woodlands, grasslands and scrub in the vicinity of deciduous trees. Most commonly associated with Ash and sallow trees but also found on a number of other deciduous trees including willow, oak, Field Maple, Alder and birch, and also on Leyland Cypress. Kidney-spot Ladybirds are not uncommon on apple trees and on herbaceous vegetation such as thistles and Nettle.

The Kidney-spot Ladybird overwinters in sheltered positions on deciduous trees, usually near the base.

Suggested survey method

Visual searching of Ash tree trunks and branches, or tree beating.

Range

Widespread through England and Wales. Few records from Scotland and absent from Ireland.

National conservation status

Widespread.

Distribution trend (1995–2015)

Stable.

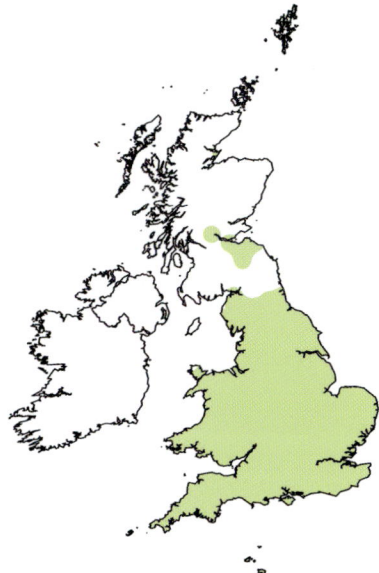

Pine Ladybird

Exochomus quadripustulatus
(Linnaeus, 1758)

The common name Pine Ladybird is not ideal for this species, as although commonly found on pine trees, it also occurs on a diverse range of deciduous trees. It is one of the first species of ladybird to emerge from overwintering, and in early spring (as early as February) it can often be seen basking on tree trunks or on fences. The front red spots have a characteristic comma shape.

Identification
Adult
Length 3–4.5mm
Background colour Black
Pattern Red spots
Number of spots 2–4 (4)
Spot fusions None
Other colour forms None
Pronotum Black
Leg colour Dark brown/black
Other features Distinct rim around the edge of the elytra

Late-instar larva Grey and spiny but with shorter bristles than either Heather Ladybird or Kidney-spot Ladybird. Characteristic white patch, appearing as a pale marking, on and around middle tubercle of first abdominal segment.

Pupa Dark and shiny with paler markings on thoracic region; larval skin encloses the entire lower part of the pupa.

Confusion species
The Pine Ladybird is commonly confused with the 2-spot Ladybird f. *quadrimaculata*, and also with the Harlequin Ladybird f. *spectabilis*. However, the first two are smaller than the Harlequin Ladybird, and the Pine Ladybird has an entirely black pronotum whereas the pronota of the Harlequin and 2-spot Ladybirds have white alongside the black markings. The Pine Ladybird has a rather domed shape and a small ridge around the elytra. The 2-spot Ladybird is more elongate.

Food
The Pine Ladybird feeds on certain scale insects, adelgids and other woolly aphids, consuming the wax covering of these insects along with the prey.

Habitat
The Pine Ladybird occupies a diverse range of habitats including deciduous, coniferous and mixed woodland, grassland, coastal habitats (cliffs and dunes), heathland and marshy areas, where it is found on a very diverse range of plants including needled conifers (particularly Scots Pine), scale-leaved conifers, Yew and many deciduous trees such as Ash, birch, sallow, willow, oak, Beech, Lime, Hazel, Sycamore, maples and Horse-chestnut. This species is very common in urban habitats, including gardens.

Confusion species			
	Pine Ladybird	2-spot Ladybird f. *quadri-maculata*	Harlequin Ladybird f. *spectabilis*
Size	3–4.5mm	4–5mm	5–8mm
Spots	Four red spots; front spots are comma-shaped	Four red spots; front spots extend to elytra edge	Four red spots
Pronotum	Black	Mainly black	White with black M-mark or solid trapezoid marking
Leg colour	Dark brown/black	Black	Brown

▲ Pine Ladybird.

72

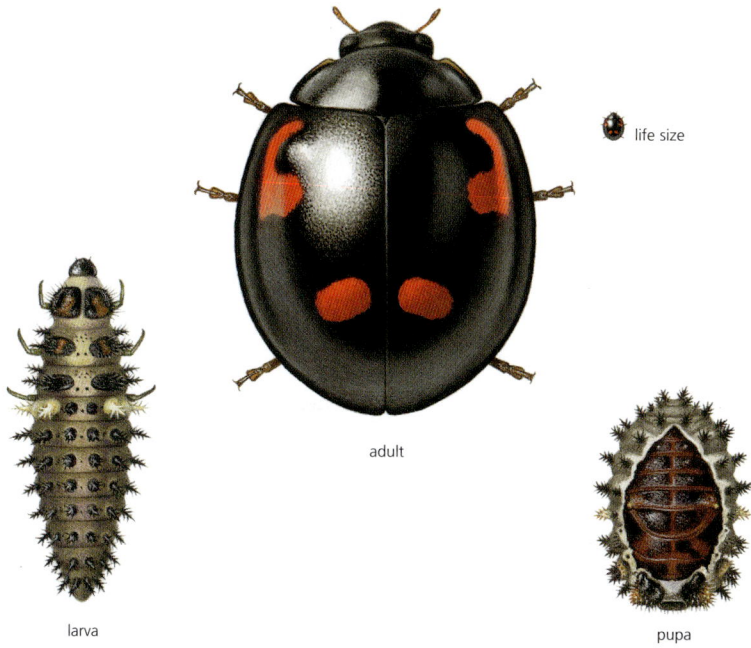

life size

adult

larva

pupa

The Pine Ladybird overwinters in leaf litter, foliage and bark crevices of evergreen trees and shrubs. Many species of ladybird spend the winter months in aggregations containing other ladybird species, and this is particularly the case with the Pine Ladybird. Indeed, the species overlaps throughout the year with a number of other species of ladybird and is commonly found with pine-specialist ladybirds but also with generalists such as 2-spot, 10-spot, 7-spot, 14-spot and Harlequin Ladybirds.

Suggested survey method
Tree beating of coniferous and deciduous trees or visual searching of trunks and branches.

Range
Widespread throughout England and Wales, and some records from Scotland.

National conservation status
Widespread.

Distribution trend (1995–2015)
Stable.

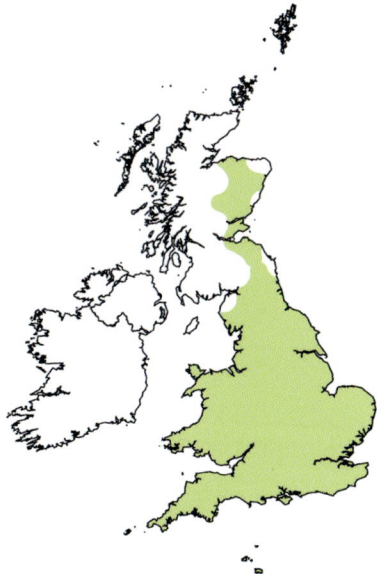

Ant-nest Ladybird

Platynaspis luteorubra
(Goeze, 1777)

Although within a tribe that comprises conspicuous ladybirds, on appearance *Platynaspis luteorubra* would perhaps be better placed alongside the inconspicuous species. One of the two ant-loving (myrmecophilous) ladybirds in Britain, with a closer tie to ants than the Scarce 7-spot Ladybird. Largely restricted to southeast England, *P. luteorubra* is found with ant species such as the Black Garden Ant (Majerus 1994). The pupa and especially the larva are unusual and very different in appearance from those of most other ladybirds, presumably because of a specialised lifestyle tied to the ants. Larvae may be found underground, feeding on subterranean aphids.

Identification
Adult
Length 2.5–3.5mm
Background colour Black
Pattern Orangey-red spots
Number of spots 4
Spot fusions None
Other colour forms None
Pronotum Black with cream anterior margin
Leg colour Reddish-brown with black femora

Other features Hairy; tibia broad, paddle-shaped; head pale (male), black (female)

Late-instar larva Very different from larvae of any other ladybirds in Britain and Ireland, including those of other species within the tribe. Yellow-grey to brown and elliptical in shape, the larva is rather flattened. Fine hairs on the body margin form a fringe. Legs not obvious.

Pupa Yellow-brown and cylindrical. Rather different from pupae of any other ladybirds in Britain and Ireland. (Not illustrated.)

Confusion species
Platynaspis luteorubra is the only one of the conspicuous ladybirds that is both black and hairy, so it is unlikely to be confused with others. It may be confused with at least two of the inconspicuous ladybirds, however, particularly *Nephus quadrimaculatus* and *Scymnus frontalis*. However, *P. luteorubra* is substantially larger than *N. quadrimaculatus* and whereas *P. luteorubra* has four red spots, *S. frontalis* has only two.

◀ *Platynaspis luteorubra.*

Confusion species			
	Platynaspis luteorubra	*Nephus quadrimaculatus*	*Scymnus frontalis*
Size	2.5–3.5mm	1.5–2mm	2.6–3.2mm
Spots	Four red spots	Four red spots	Two red spots

adult

life size

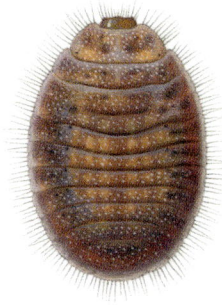

larva

Food
Platynaspis luteorubra is a predatory ladybird that feeds on aphids.

Habitat
Platynaspis luteorubra is a species of dry sandy and chalky habitats occupied by ants. It may be found in low-growing vegetation in association with ants. There is one recent record of the species beaten from a pine tree.
   The overwintering sites are unknown.

Suggested survey method
Sweep netting or visual searching of appropriate habitats with ants.

Range
Largely restricted to southeast England.

National conservation status
Very local.

Distribution trend (1995–2015)
Insufficient data.

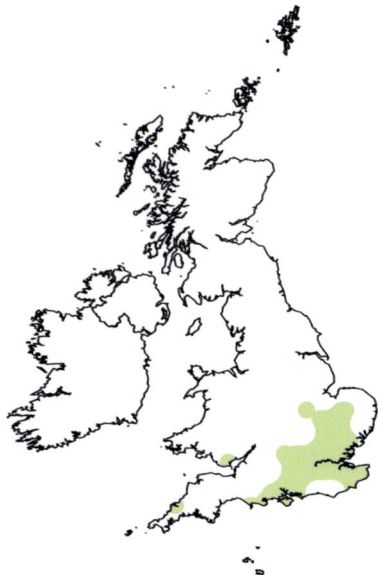

Orange Ladybird

Halyzia sedecimguttata
(Linnaeus, 1758)

Bright orange ladybird with white spots. Historically considered an indicator of ancient woodland but has seen a dramatic expansion into other habitats since late 1980s.

Identification
Adult
Length 4.5–6mm
Background colour Orange
Pattern White spots
Number of spots 12–16 (16)
Spot fusions None
Other colour forms None
Pronotum Orange with translucent edging
Leg colour Orange

Late-instar larva Pale cream with bright yellow longitudinal stripes running between inner and middle rows of black tubercles. The first thoracic segment is bright yellow, head and legs are pale cream, and the body has a smooth surface.

▲ Orange Ladybird.

Pupa Highly distinctive; black with pairs of bright yellow spots on the lateral edges of abdominal segments one and two, and duller ones on segments five, six and seven.

Confusion species
Orange Ladybird adults are often confused with Cream-spot Ladybirds, although the former are usually bright orange and the latter maroon-brown. Orange Ladybird larva can be distinguished from 22-spot Ladybird larva by colour, longer legs and pale head of the former, and middle tubercle on each side of first abdominal segment having black tip (yellow in 22-spot Ladybird).

Food
Larvae and adults feed on powdery white mildews growing on the leaves of deciduous trees.

Habitat
The Orange Ladybird is a woodland species but is increasingly common on urban trees. Sycamore, oak and Ash are favoured, although in recent years an increasing number of adults and larvae have been found on Hawthorn. This species is also recorded from a number of other deciduous trees and shrubs including Dogwood, Lime, Hazel, sallow and birch, particularly where Ivy is present. There is a scattering of records from coniferous trees including Douglas Fir and Scots Pine. Feeding has only been observed on Sycamore, Ash, birch, Dogwood and Field Maple, and all other plant species are yet to be proven as hosts.

Confusion species		
	Orange Ladybird	**Cream-spot Ladybird**
Size	4.5–6mm	4–5mm
Background colour	Orange	Maroon-brown
Spots	16 white spots appearing more scattered across the elytra than in the Cream-spot Ladybird	14 distinct round cream spots often appearing in ordered transverse rows (1–3–2–1 from front to rear of each elytron)
Pronotum	Orange with translucent edging	Maroon with lateral cream markings
Leg colour	Orange	Brown

 life size

adult

larva

pupa

The Orange Ladybird overwinters in sheltered spots on the bark of deciduous and coniferous trees. It has also been found on Rhododendron and Ivy and, in particularly harsh winters, in the litter layer.

Suggested survey method
Tree beating or visual searching of deciduous trees. This species is attracted to light and is commonly found in moth traps.

Range
Widely distributed across Britain and Ireland.

National conservation status
Very widespread.

Distribution trend (1995–2015)
Increasing.

22-spot Ladybird — *Psyllobora vigintiduopunctata*
(Linnaeus, 1758)

Small ladybird with lemon-yellow elytra and pronotum. The elytra have 20–22 small black spots (rarely fused) and the pronotum a further five. The only ladybird species in Britain and Ireland in which the larva, pupa and adult are all of a comparable colour and pattern.

Identification
Adult
Length 3–4mm
Background colour Bright lemon-yellow
Pattern Black spots
Number of spots 20–22 (22)
Spot fusions Rare
Other colour forms None
Pronotum Yellow or white with four discrete black spots in a semi-circle and a black triangle at the mid-base
Leg colour Brown

▲ 22-spot Ladybird.

Late-instar larva Yellow with black tubercles and black head, distinguished from Orange Ladybird larva by dark head of 22-spot Ladybird and middle tubercle on each side of first abdominal segment yellow (black-tipped in Orange Ladybird).

Pupa Yellow with black spots.

Confusion species
14-spot and 16-spot Ladybirds are sometimes confused with 22-spot Ladybirds. The 14-spot Ladybird is generally less bright yellow than the 22-spot Ladybird, but additionally the black spots of the 14-spot are larger, usually more rectangular in shape and often fused together. The pronota of the two species are also distinctive. The 16-spot Ladybird is notably smaller than the 22-spot Ladybird, but is also paler and duller in colour.

Food
The 22-spot Ladybird feeds on mildew that develops on the upper surfaces of the leaves of Hogweed and other umbellifers. Towards the end of the season, when Hogweed is senescing, this species will feed on the mildews of other herbaceous plants, including Creeping Thistle, and young trees, especially oak.

Habitat
The 22-spot Ladybird generally favours low vegetation in grassland habitats, and is often reported from roadside and field-side

Confusion species			
	22-spot Ladybird	**14-spot Ladybird**	**16-spot Ladybird**
Size	3–4mm	3.5–4.5mm	3mm
Background colour	Bright yellow	Yellow	Beige
Spots	Elytra have 20–22 small black spots (rarely fused) and the pronotum a further five	Usually 14 rectangular spots often fused	Usually 16 spots, with three lateral spots on each elytron usually fused
Pronotum	Yellow with four discrete black spots in a semi-circle and a black triangle at the mid-base	Yellow or cream, usually with crown-shaped mark	Beige with black spots

life size

adult

larva

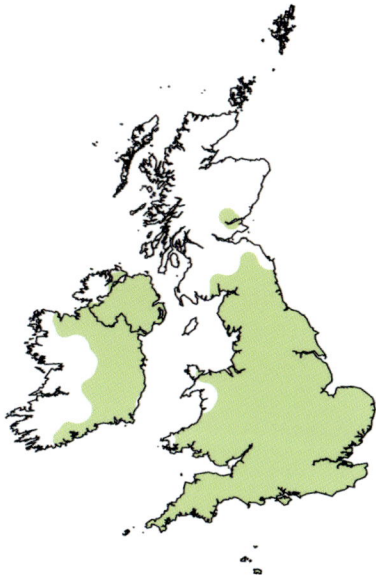

pupa

vegetation, usually on low herbs, although it will visit the lower branches of young trees. It is occasionally found in woodlands and in coastal habitats such as sand dunes. The 22-spot Ladybird is often found on Hogweed, but many records are associated with other umbellifers, such as Cow Parsley, and it has also been found on Teasel, Foxglove, Mugwort, burdock and Wild Parsnip, and occasionally on shrubs and trees including Ash, willow, sallow, Hornbeam, Hawthorn and oak.

The 22-spot Ladybird overwinters in herbage, including grasses, close to the ground, or sometimes in Ivy.

Suggested survey method
Sweep netting or visual searching of herbaceous plants with mildew, especially Hogweed.

Range
Very widespread species in England, Wales and Ireland. Limited distribution in Scotland.

National conservation status
Very widespread.

Distribution trend (1995–2015)
Decreasing.

Water Ladybird   *Anisosticta novemdecimpunctata*
(Linnaeus, 1758)

In the winter Water Ladybirds shelter between the sheaths of dead Reed and Reedmace leaves, and at this time are buff-coloured, usually with 19 black spots. In the spring they disperse to feed on aphids on living Reed and the background colour of their elytra rapidly changes from buff to red warning colours.

Identification
Adult
Length 4mm
Background colour Buff/beige (July–April), reddish (April–June)
Pattern Black spots
Number of spots 15–21 (19)
Spot fusions Sometimes
Other colour forms None
Pronotum Buff/beige with six black spots; rounded at the sides with greatest width in the middle
Leg colour Pale brown
Other features Distinctly elongate and flattened in shape

Late-instar larva Thoracic region cream/white with dark patches; abdomen pale grey with rows of black tubercles running longitudinally; fine hairs projecting from tubercles.

Pupa Dark with pale brown and orange markings. Small ragged teeth along the sides.

Food
The Water Ladybird feeds on aphids found on Reed, Reedmace and Reed Sweet-grass, growing alongside lakes and rivers.

Habitat
The Water Ladybird is a habitat specialist, occupying reedbeds and grassland in marshy or wet habitats. This species can often be found on the emergent vegetation (Reed, Reedmace, Reed Sweet-grass and rushes) surrounding ponds, but there are also a few records from exposed riverine sediments.

80

▲ Water Ladybird.

life size

81

larva

pupa

adult

form

The Water Ladybird overwinters between leaves and in stems of Reed and Reedmace, and in grass tussocks.

Suggested survey method
Sweep netting of Reed and wetland grasses. Additionally, visual searching the stems of Reedmace and Reed for wintering adults.

Range
Widespread through England and Wales. A few recent records from Scotland. Absent from Ireland (but there are two old records).

National conservation status
Widespread.

Distribution trend (1995–2015)
Decreasing.

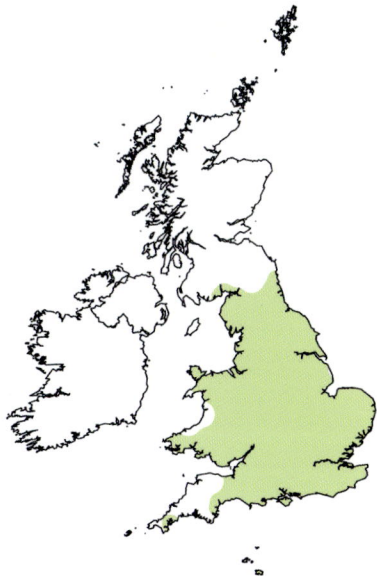

16-spot Ladybird

Tytthaspis sedecimpunctata
(Linnaeus, 1761)

Small size, camouflaged colouration and a tendency to stay close to the ground make this a species that is very easy to overlook, although it can be abundant in grassland. A central black line between the elytra and the distinctive spot pattern, with two inner lines of three unfused spots and two outer lines of five spots, three of which are usually fused to form a zigzag mark, are diagnostic.

Identification
Adult
Length 3mm
Background colour Beige
Pattern Black spots
Number of spots 13–18 (16)
Spot fusions Common; the three lateral spots on each elytron are usually fused
Other colour forms Melanic (black) forms – rare
Pronotum Beige with black spots
Leg colour Brown

Late-instar larva Pale brown-grey with fine long black hairs, giving bristled appearance, emanating from darker coloured tubercles.

Pupa Pale yellowish-brown with four rows of diffuse darker markings running longitudinally; remains of hairy fourth-instar larval skin visible at base.

Food
The 16-spot Ladybird is one of the few non-predatory ladybirds; it feeds on pollen, nectar and fungi.

Habitat
Abundant in grassland but also in scrub habitats, saltmarsh and dune systems. Mostly found on grasses and other low plants, including Reed, Nettle, Dandelion, knapweed, Hogweed, Cow Parsley and buttercup. This species has also been recorded from shrubs, including Gorse, and from Scots Pine.

82

▲ 16-spot Ladybird.

life size

adult

larva

83

pupa

The 16-spot Ladybird overwinters in low herbage, on Gorse, in plant litter, on logs, fence posts and stone walls, often in extremely large aggregations (hundreds or thousands of individuals).

Suggested survey method
Sweep netting in grassland habitats.

Range
Widespread in England and sparsely distributed in Wales. Largely absent from Scotland and absent from Ireland.

National conservation status
Widespread.

Distribution trend (1995–2015)
Stable.

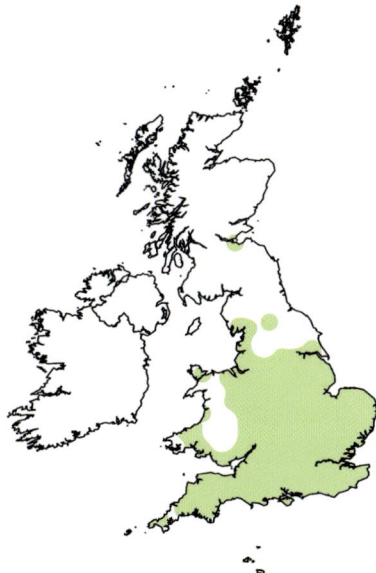

13-spot Ladybird

Hippodamia tredecimpunctata
(Linnaeus, 1758)

An elongate and flattened ladybird with distinctive pronotum marking. Until relatively recently there had not been any records of the 13-spot Ladybird in Britain since 1952, but over the last 10 or so years there have been various sightings, mainly from the south of England, suggesting that the species is a migrant from mainland Europe that is turning up more often than in the past.

Identification
Adult
Length 5–7mm
Background colour Orange-red
Pattern Black spots
Number of spots 7–15 (13)
Spot fusions Rare
Other colour forms Melanic (black) forms – rare
Pronotum White with distinctive black pattern
Leg colour Black
Other features Distinctly elongate and slightly flattened

Late-instar larva Dark grey with six rows of tubercles bearing short stubby hairs. The tubercles are inky black except for those on abdominal segment four, which are all pale grey, as are the middle and outer tubercles on the first abdominal segment.

Pupa Dark grey with two rows of small black spots running longitudinally (formed from inner tubercles). Pale lateral patches on first abdominal segment. (Not illustrated.)

Food
The 13-spot Ladybird is a predatory ladybird that feeds on aphids.

Habitat
The majority of records in Britain are from Reed and grasses in marshy lowland habitats such as fens, marshes, riverbanks and dune systems. In continental Europe, the 13-spot Ladybird is mainly found on plants in marshy habitats but is also commonly found on crops. Indeed, many studies on agricultural systems in the Palaearctic region list the 13-spot Ladybird as one of the three most abundant species of ladybird, particularly favouring dense and humid stands.

▲ 13-spot Ladybird.

life size

adult

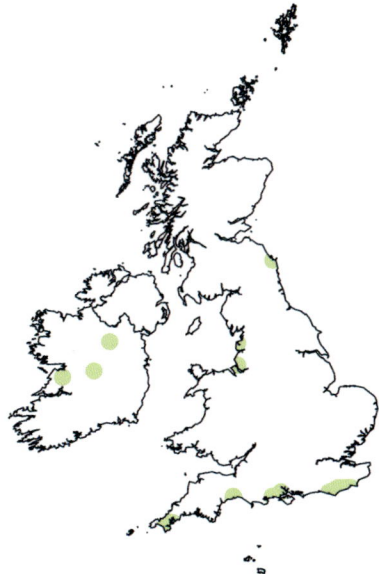

larva

Overwintering sites are unknown in Britain, but in continental Europe the 13-spot Ladybird favours litter or upper soil layers in damp habitats.

Suggested survey method
Sweep netting of Reed and grasses in damp habitats.

Range
Occasional records from the south of England and the Channel Islands but very rare (though common across continental Europe). Over the last 10 years there has been an increase in the number of sightings. There appears to be an established breeding colony in Sussex, and in recent years larvae have also been recorded in south Devon. There are several recent records from the Channel Islands. There was also an isolated recent report of the species in Northumberland.

National conservation status
Very local.

Distribution trend (1995–2015)
Insufficient data.

Adonis Ladybird   *Hippodamia variegata* (Goeze, 1777)

A medium ladybird with an elongate body. Most commonly with seven spots, although the number varies; a distinctive feature is that usually the spots are positioned in the posterior half of the elytra. Those spots nearer the front, if present, are usually small.

Identification
Adult
Length 4–5mm
Background colour Red
Pattern Black spots
Number of spots 3–15 (7)
Spot fusions Common
Other colour forms None
Pronotum White with distinctive black pattern with an undulating edge; undulations often fuse to leave white spots
Leg colour Black

Late-instar larva Greyish-brown with a pale orange transverse band at the back of the head in front of the first thoracic segment and a thicker, orange transverse band between the first and second thoracic segment. Outer and middle tubercles on first abdominal segment are orange; all other tubercles dark grey/black.

▲ Adonis Ladybird.

Pupa Distinct dark rectangular markings run longitudinally along the pale orange pupa, which has remains of shed larval skin, with fine hairs, visible at the base.

Confusion species
The Adonis Ladybird is sometimes confused with the 11-spot Ladybird, which may occur in the same dry habitats, but the pronotum pattern of the Adonis Ladybird is distinctive.

Food
The Adonis Ladybird is a predatory ladybird that feeds on aphids.

Habitat
Often a coastal species, the Adonis Ladybird inhabits dune systems, but is increasingly also being found in many inland areas. It favours habitats where the vegetation surrounds, or is adjacent to, areas of dry soil, shingle or shale, and is also found on waste ground and at gravel pits, often walking across bare ground. This species is one of the most abundant ladybirds in Mediterranean Europe, where it can be very common in fields, especially where there is bare ground and weedy vegetation. It has been reported on a number of herbaceous plants including Nettle, Wild Parsnip, thistles, Cow Parsley, Hogweed and Tansy.

The Adonis Ladybird overwinters in leaf litter and on low plants.

Suggested survey method
Sweep netting or visual searching of herbaceous plants, particularly weedy ground with bare patches in warm and sheltered locations.

Confusion species		
	Adonis Ladybird	**11-spot Ladybird**
Size	4–5mm	4–5mm
Spots	Usually seven, with six commonly placed at the rear of the elytra	Usually 11
Pronotum	White with distinctive undulating-edged black pattern; undulations often fuse to leave white spots	Black with anterior-lateral white marks; broadest at base

life size

adult

larva

pupa

forms

Range
Britain is at the northwestern edge of the
range, but in recent years the species has
been increasing, possibly aided by climate
warming. Currently distributed in southern,
eastern and central England as far north as
Yorkshire, with few records from Wales and
Scotland but none from Ireland.

National conservation status
Local.

Distribution trend (1995–2015)
Stable.

Larch Ladybird   *Aphidecta obliterata* (Linnaeus, 1758)

Dull appearance compared to other conspicuous ladybirds. Light brown elytra, usually with a diffuse dark line running down the midsection, and sometimes speckled with tiny brown spots. Black-brown M-shaped mark on pronotum.

Identification
Adult
Length 4–5mm
Background colour Light brown, sometimes speckled with tiny brown spots
Pattern None, or dark oblique line posteriorly
Number of spots 0–10 (0)
Spot fusions Rare
Other colour forms Melanic (black) forms – rare in Britain and Ireland
Pronotum Pale with black-brown M-mark or spots
Leg colour Brown
Other features Can appear well camouflaged against its background

Late-instar larva Light grey; thoracic region with darker speckled patches; a pair of orange-yellow spots on middle and outer tubercles of first abdominal segment, remaining tubercles dark grey with short stubby hairs protruding. Legs uniformly dark. Very similar to larva of 18-spot Ladybird (but note leg colour difference), so rearing through to adulthood is recommended for accurate identification.

Pupa Pale grey with four rows of dark grey squarish patches running longitudinally. The anterior section has dark zigzag markings, and there are yellow patches on the sides of the first abdominal segment.

Food
Primarily adelgids but also aphids and scale insects.

Habitat
The Larch Ladybird is a conifer specialist, commonly found in coniferous or mixed woodlands. Most records are from larch trees, but it is also recorded from Norway Spruce, Douglas Fir and occasionally Scots Pine.

▲ Larch Ladybird.

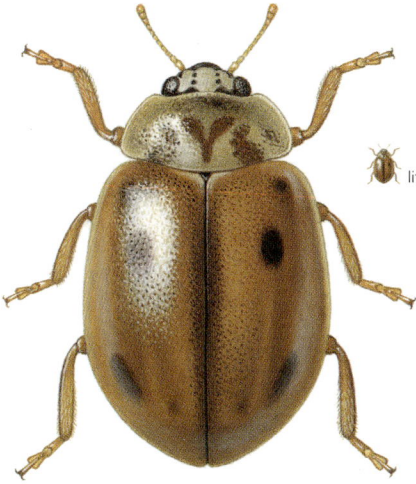

life size

adult

larva

pupa

form

The Larch Ladybird overwinters primarily in bark crevices on larch, Norway Spruce and Douglas Fir; it is not uncommon for this species to be beaten from Ivy growing on or near these trees.

Suggested survey method
Tree beating or visual searching of larch trees.

Range
Widespread species in Britain and Ireland.

National conservation status
Widespread.

Distribution trend (1995–2015)
Decreasing.

2-spot Ladybird *Adalia bipunctata* (Linnaeus, 1758)

Highly variable colour pattern (polymorphic). Typical form (red background colour with a single black spot on each elytron) is by far the most common, but there are many permutations, some with a black background colour to the elytra.

Identification

Adult
Length 4–5mm
Background colour (1) 'Typical' form (*typica*) and (2) f. *annulata*: red; (3) 'four-spot melanic' (*quadrimaculata*) and (4) 'six-spot melanic' (*sexpustulata*): black
Pattern (1) With two black spots; (2) with two to six irregular elongate black spots; (3) with four red spots; (4) with six red spots
Number of spots 0–16 (2)
Spot fusions Sometimes
Other colour forms Many and common
Pronotum White with black spots, a black M-mark or mainly black
Leg colour Black (a good feature for distinguishing from 10-spot Ladybird, which has brown legs)

▲ 2-spot Ladybird f. *typica*.

Other features Very variable in colour and pattern

Late-instar larva Variable and closely resembles larva of 10-spot Ladybird; 2-spot Ladybird larva is dark grey (10-spot larva is pale grey) with a triangle of orange spots across the first and fourth abdominal segments. Fine hairs project from the tubercles, and the outer tubercles on abdominal segments five to eight are dark (pale in 10-spot).

Pupa Closely resembles pupa of 10-spot Ladybird; 2-spot pupa has a black front section but is otherwise cream with six rows of dark spots running longitudinally (10-spot pupa is similar but with a paler overall appearance, two orange spots on edges of first abdominal segment and an orange patch in the middle of segments four to six).

Confusion species

The typical red form of the 2-spot Ladybird is quite distinctive. However, this species is very variable, with many other colour forms. The 10-spot Ladybird is similar in size and appearance, so leg colour can be a useful identifying feature: the 2-spot Ladybird has black legs, whereas the 10-spot Ladybird has brown legs. Black forms of the 2-spot Ladybird may be confused with the Pine Ladybird or with black forms of 10-spot and Harlequin Ladybirds.

Food

The 2-spot Ladybird is a predatory ladybird that feeds on aphids.

Confusion species				
	2-spot Ladybird f. *quadrimaculata*	**10-spot Ladybird f. *bimaculata***	**Pine Ladybird**	**Harlequin Ladybird f. *spectabilis***
Size	4–5mm	3.5–4.5mm	3–4.5mm	5–8mm
Spots	Four red spots; front spots extend to elytra edge	Two red spots at front	Four red spots; front spots are comma-shaped	Four red spots
Pronotum	Mainly black	Black with white edges	Black	White with black M-mark or solid trapezoid marking
Leg colour	Black	Brown	Dark brown/black	Brown

larvae

adult
f. *typica*

pupa

life size

f. *quadrimaculata*

f. *sexpustulata*

forms

f. *bar annulata*

f. *intermediate annulata*

▲ 2-spot Ladybird f. *quadrimaculata*.

▲ 2-spot Ladybirds mating: f. *typica* female and f. *sexpustulata* male.

Habitat

The 2-spot Ladybird can be found in a variety of habitats. Many records are from urban areas where deciduous trees are abundant. The species also occurs in mature woodlands (both deciduous and coniferous), scrub and grasslands, orchards and crops, including cereals but particularly broad-leaved crops such as Field Bean. There are a number of records from wetlands and coastal habitats, particularly dune systems. The 2-spot Ladybird is readily found on mature Lime or Sycamore trees, for example in parks or churchyards, and there are many observations from herbaceous plants such as Nettle, thistles, Rosebay Willowherb, Fat-hen and ornamental plants such as roses, Buddleja, Lavender and hebe. The 2-spot Ladybird overlaps with a number of other species of ladybird and is commonly found with 10-spot, 7-spot, 14-spot, Pine and Harlequin Ladybirds.

It overwinters in locations at slightly elevated positions, such as in the attics and upstairs window-frames of houses, or on tree trunks and under bark.

Suggested survey method

Tree beating or visual searching of deciduous trees.

Range

Widely distributed across Britain and Ireland, but absent from northern Scotland.

National conservation status

Ubiquitous.

Distribution trend (1995–2015)

Decreasing.

▲ 2-spot Ladybirds f. *sexpustulata* overwintering.

10-spot Ladybird

Adalia decempunctata
(Linnaeus, 1758)

The most variable ladybird in Britain and Ireland in terms of colour and pattern.

Identification
Adult
Length 3.5–4.5mm
Background colour (1) 'Typical' form (*decempunctata*): yellow, orange or red; (2) 'chequered' form (*decempustulata*): brown or black; (3) 'melanic' form (*bimaculata*): purple, dark brown or black
Pattern (1) With 0–15 maroon, dark brown or black spots; (2) with cream, yellow, red or light-brown grid-like markings giving a chequered pattern; (3) with two yellow, orange or red shoulder flashes
Number of spots 0–15 (10)
Spot fusions Common
Other colour forms Various and common
Pronotum White with five dark spots, which may be fused, or dark trapezium mark
Leg colour Brown (a good feature for distinguishing from 2-spot Ladybird, which has black legs)

Other features Extremely variable in colour and pattern

Late-instar larva Closely resembles larva of 2-spot Ladybird; 10-spot Ladybird larva is pale grey (2-spot larva is dark grey), with a triangle of yellow spots across the first and fourth abdominal segments. There are fine hairs projecting from tubercles, and the outer tubercles on abdominal segments five to eight are pale (dark in 2-spot).

Pupa Closely resembles pupa of 2-spot Ladybird; 10-spot Ladybird pupa has a black front section but is otherwise cream with six rows of dark spots running longitudinally. There are two orange spots on the edges of the first abdominal segment and an orange patch in the middle of abdominal segments four to six (2-spot Ladybird pupa is similar but with a darker overall appearance and lacking the orange markings).

93

◀ 10-spot Ladybird
f. *decempunctata*.

Confusion species			
	10-spot Ladybird f. *decempunctata*	2-spot Ladybird f. *typica*	Harlequin Ladybird f. *succinea*
Size	3.5–4.5mm	4–5mm	5–8mm
Spots	0–15 black spots; usually one 'shoulder' spot on each elytron	Two black spots	0–21 black spots; usually two 'shoulder' spots on each elytron
Pronotum	White with black spots	White with black M-mark	White with black M-mark or solid trapezoid marking
Leg colour	Brown	Black	Brown

life size

94

larva

adult
f. *decempunctata*

pupa

f. *decempunctata*

f. *decempunctata*

f. *decempustulata*

f. *bimaculata*

Food

The 10-spot Ladybird is a predatory ladybird that feeds on aphids.

Habitat

The 10-spot Ladybird is more habitat-specific than the 2-spot Ladybird but still occupies a variety of habitats. It is generally found on deciduous trees and hedgerows in woodland and urban areas. Like the 2-spot Ladybird, this species can be readily found on mature Lime or Sycamore trees, for example in parks or churchyards, but also oak, willow, Hawthorn and Blackthorn. Many records are from deciduous and conifer woodlands, scrub and grasslands, and less commonly from wetlands and coastal habitats, particularly dune systems. Many observations are from herbaceous

▲ 10-spot Ladybird f. *decempunctata*.

plants such as Nettle, Hogweed, thistles, Rosebay Willowherb, Fat-hen and garden plants such as roses and Buddleja, fruit trees and shrubs such as Blackcurrant and cherry, but less commonly than 2-spot Ladybird. The 10-spot Ladybird overlaps with a number of other species of ladybird and is commonly found with 2-spot, 7-spot, 14-spot, Pine and Harlequin Ladybirds.

It overwinters in leaf litter, plant debris and Beech nuts.

Suggested survey method
Tree beating or visual searching of deciduous trees and hedgerows.

Range
Widely distributed across Britain and Ireland, although absent from northern Scotland.

National conservation status
Ubiquitous.

Distribution trend (1995–2015)
Stable.

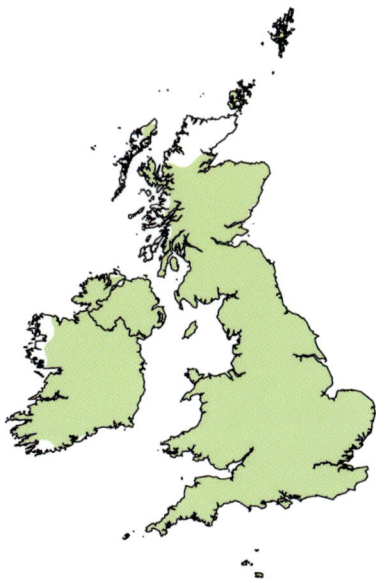

Hieroglyphic Ladybird

Coccinella hieroglyphica
Linnaeus, 1758

Variable and elusive species, coloured anything from bronzy-brown with a squiggly black mark resembling an Egyptian hieroglyph, to completely black. This heathland specialist is thought to feed on leaf (chrysomelid) beetles. In Britain and Ireland this includes the Heather Leaf Beetle, but in continental Europe the Hieroglyphic Ladybird is reported to feed on aphids. It is perhaps the restricted diet of this species in Britain and Ireland that limits its abundance.

Identification
Adult
Length 4–5mm
Background colour Brown or black
Pattern Black stripes, spots and patches, sometimes resembling an Egyptian hieroglyph
Number of spots 0–7 (5)
Spot fusions Common
Other colour forms Melanic (black) forms – common
Pronotum Black with anterior-lateral white marks
Leg colour Black

Late-instar larva Dark grey/black, with black tubercles producing fine hairs. The middle and outer tubercles on abdominal segments one and four are pale yellow/whitish. There are pale yellow patches in the centre of the second and third thoracic segments.

Pupa Orange with dark spots, slight sculpturing to outer edge. (Not illustrated.)

Food
The Hieroglyphic Ladybird larva preys on the larvae of the Heather Leaf Beetle, although the adult is also known to consume the Heather Aphid. Also, larvae of chrysomelid beetles within the genera *Altica* and *Galerucella*.

Habitat
The Hieroglyphic Ladybird is usually found in heathland, often on old heather plants, or where the heathland has been invaded by scrub. This species is also found on acid grassland and heathland mosaics.

It overwinters in litter under heather, pine trees and Gorse bushes.

96

▲ Hieroglyphic Ladybird.

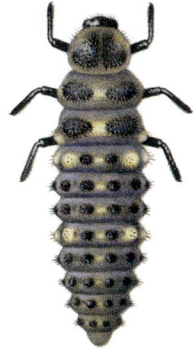

life size

97

adult

larva

forms

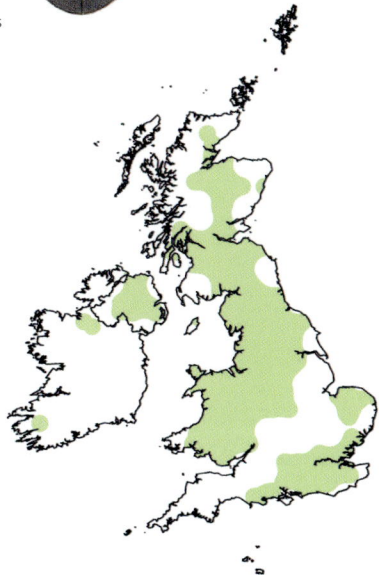

Suggested survey method
Sweep netting of heather; overwintering individuals may be found by beating pine trees or Gorse bushes growing in the vicinity of heather heathland.

Range
Widespread in Britain and Ireland.

National conservation status
Very local.

Distribution trend (1995–2015)
Decreasing.

Scarce 7-spot Ladybird

Coccinella magnifica
Redtenbacher, 1843

The Scarce 7-spot Ladybird is very similar to the 7-spot Ladybird, although in the former often the central spots are slightly larger than the outer spots. However, the only way to confirm the identification is to check the underside for the four white marks by the legs. It is ant-loving (myrmecophilous) and in Britain only found living close to the nests of wood ants of the genus *Formica*, most commonly the Red Wood Ant.

Identification
Adult
Length 6–8mm
Background colour Red
Pattern Black spots, central spots comparatively large and foremost spots comparatively small
Number of spots 5–11 (7)
Spot fusions Rare
Other colour forms None
Pronotum Black with anterior-lateral white marks
Leg colour Black
Other features Presence of four small white triangular marks on the underside

Late-instar larva Dark grey/brown, with black tubercles producing fine hairs. There are pale yellow lateral patches on the first thoracic segment; sides of second and third thoracic segments are pale grey and yellow respectively. The middle and outer tubercles on abdominal segments one and four are yellow.

Pupa Often pale orange with two rows of dark triangular markings running down the middle, and dark spots on anterior section, but potentially more diffuse than 7-spot Ladybird.

Confusion species
Very similar to the 7-spot Ladybird, but the Scarce 7-spot Ladybird usually has a more domed shape and larger spots. The best way to distinguish the two species is to look at the white markings below the legs on their undersides; the Scarce 7-spot Ladybird has four white markings whereas the 7-spot Ladybird only has two.

Food
The Scarce 7-spot Ladybird is a predatory ladybird that feeds on aphids.

Confusion species		
	Scarce 7-spot Ladybird	7-spot Ladybird
Underside	Four white triangular markings	Two white triangular markings

▲ Scarce 7-spot Ladybird.

life size

adult

larva

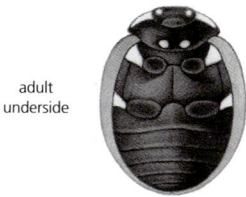

adult
underside

pupa

Habitat

The Scarce 7-spot Ladybird is found in habitats close to (although never in) wood ant nests, usually in woodland but also in heathland. It is recorded from various plants including Gorse, thistles and heathers, but is most commonly found on Scots Pine.

The Scarce 7-spot Ladybird overwinters in various locations but always within a few metres of wood ant nests.

Suggested survey method

Tree beating or visual searching of Scots Pine trees and other plants in heathland, close to wood ants.

Range

Largely restricted to southeast England, with a few more northerly records.

National conservation status

Very local.

Distribution trend (1995–2015)

Stable.

5-spot Ladybird

Coccinella quinquepunctata
Linnaeus, 1758

The 5-spot Ladybird is a red ladybird with five black spots. Very localised in Britain and absent from Ireland. Rarely found more than a few metres from unstable river shingle, but common in many habitats in continental Europe.

Identification
Adult
Length 4–5mm
Background colour Red
Pattern Black spots
Number of spots 5–9 (5)
Spot fusions Very rare
Other colour forms None
Pronotum Black with anterior-lateral white marks
Leg colour Black
Other features Quite rounded and domed in shape

Late-instar larva Dark grey, with black tubercles producing fine hairs. Lateral patches on the first thoracic segment are bright orange, as are the middle and outer tubercles on abdominal segments one and four, and the outer tubercle on segments six and seven.

Pupa Very dark, with orange spots on the lateral edges of abdominal segment one.

Food
The 5-spot Ladybird is a predatory ladybird that feeds on aphids.

Habitat
In Britain the 5-spot Ladybird is found only on unstable river shingle, where it is reported on Nettle, thistles, Bitter-cress, docks and Angelica. There are a few records of this species on Broom. In continental Europe the 5-spot Ladybird occurs in more varied habitats.

It overwinters on Gorse, under shingle stones and in leaf litter.

Suggested survey method
Visual searching of herbaceous plants on river shingle.

▲ 5-spot Ladybird.

life size

adult

larva

pupa

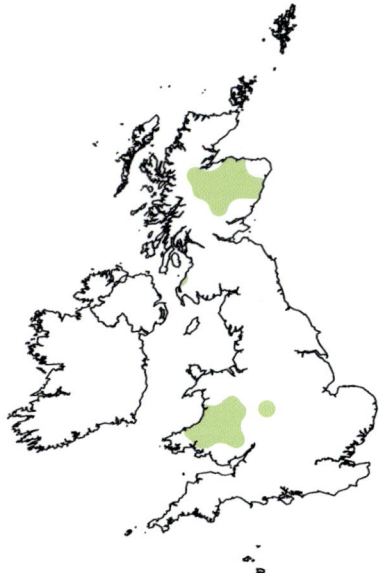

Range
Very local, with few records from Britain, but the Afon Ystwyth, Afon Rheidol and River Severn in Wales, and the River Spey in Scotland, are particular hotspots. A few recent records in England, close to the Welsh border. Not recorded in Ireland.

National conservation status
Very local.

Distribution trend (1995–2015)
Stable.

7-spot Ladybird

Coccinella septempunctata
Linnaeus, 1758

An iconic ladybird; red with seven black spots.

Identification
Adult
Length 5–8mm
Background colour Red
Pattern Black spots
Number of spots 0–9 (7)
Spot fusions Very rare
Other colour forms Melanic (black) forms – very rare
Pronotum Black with anterior-lateral white marks
Leg colour Black
Other features Can be distinguished from Scarce 7-spot Ladybird by presence of two small white triangular marks on the underside (Scarce 7-spot Ladybird has an additional pair of white marks)

Late-instar larva Dark grey/black with bluish tinge and black tubercles producing fine hairs. There are dark orange lateral patches on the first thoracic segment, the sides of the second and third thoracic segments are dark grey/black, and the middle and outer tubercles on abdominal segments one and four are dark orange.

Pupa Often pale orange with two rows of dark triangular markings running down the middle, and four small dark spots on anterior section.

Confusion species
Very similar to the Scarce 7-spot Ladybird, but the 7-spot Ladybird usually has smaller spots than the Scarce 7-spot. The clearest way to distinguish the two species is to look at the white markings below the legs on their undersides; the Scarce 7-spot Ladybird has four white markings, whereas the 7-spot Ladybird only has two.

Food
The 7-spot Ladybird is a predatory ladybird that feeds on aphids.

Habitat
The 7-spot Ladybird occurs in many habitats including dunes, grassland, heathland, scrub, coniferous, deciduous and mixed woodland – usually on low herbage. It is very common in gardens and agricultural habitats (both cereals and broad-leaved crops such as Field Bean) as well as Hawthorn hedgerows and occasionally on trees (including oak, Lime, Sycamore, Douglas Fir and Scots Pine) and on Ivy growing

Confusion species		
	7-spot Ladybird	Scarce 7-spot Ladybird
Underside	Two white triangular markings	Four white triangular markings

◀ 7-spot Ladybird.

102

life size

adult

larva

adult
underside

pupa

around the trunks. Plant records include Nettle, Rosebay Willowherb, thistles, Cow Parsley, Wild Carrot, Yarrow, ragwort, heathers, Angelica, Hogweed, Fat-hen, Reed, dead-nettles and Mugwort, as well as numerous garden plants such as roses, Lavender, Buddleja, Wallflower, Peony and Camellia.

The 7-spot Ladybird overwinters in a variety of places including low herbage, Gorse, conifer foliage and in leaf litter, often in curled dead leaves. In the autumn this species is commonly seen sheltering in senescing seed heads, such as those forming on Rosebay Willowherb, Teasel and Cow Parsley.

Suggested survey method
Sweep netting or visual searching of herbaceous plants.

Range
Widely distributed across Britain and Ireland.

National conservation status
Ubiquitous.

Distribution trend (1995–2015)
Decreasing.

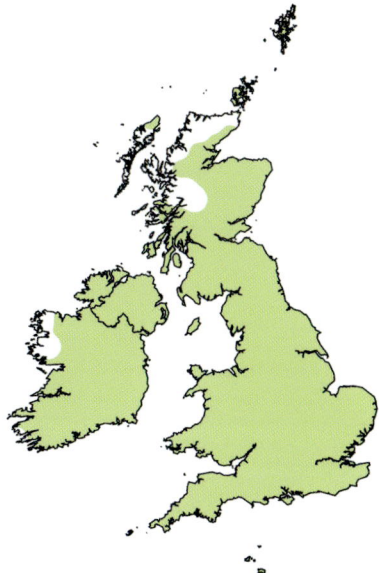

11-spot Ladybird

Coccinella undecimpunctata
Linnaeus, 1758

A medium and slightly elongate ladybird. Some specimens bear diffuse yellow rings around their 11 black spots (in a similar way to the Eyed Ladybird).

Identification
Adult
Length 4–5mm
Background colour Red
Pattern Black spots
Number of spots 7–11 (11)
Spot fusions Uncommon
Other colour forms None
Pronotum Black with anterior-lateral white marks; broadest at base
Leg colour Black
Other features Black spots occasionally surrounded by a thin yellow ring

Late-instar larva Closely resembles 7-spot Ladybird larva, but smaller and without the conspicuous orange lateral patches on the first thoracic segment. Abdomen has orange spots in pairs on a grey-black background.

Pupa Black front section but otherwise cream with inner tubercles on abdominal segments forming two dark bands running longitudinally. There are orange lateral patches on the first abdominal segment; the inner and outer tubercles on the fourth abdominal segment are also orange.

Confusion species
The 11-spot Ladybird is sometimes confused with the Adonis Ladybird, which may occur in the same dry habitats, but the pronotum pattern of the Adonis Ladybird is distinctive.

Food
The 11-spot Ladybird is a predatory ladybird that feeds on aphids.

Habitat
The 11-spot Ladybird is most often found in herbaceous vegetation, particularly dune systems and coastal habitats, or in inland regions with sandy soils. It is commonly associated with Sea Radish, Nettle, Gorse, Rosebay Willowherb and thistles. There is a scattering of records from deciduous trees including Ash, Beech, Sycamore and oak.

It overwinters in leaf litter and buildings.

Suggested survey method
Sweep netting or visual searching of herbaceous plants, or beating of scrub, particularly Gorse, in dune systems.

Confusion species		
	11-spot Ladybird	**Adonis Ladybird**
Size	4–5mm	4–5mm
Spots	Usually 11	Usually seven, with six commonly placed at the rear of the elytra
Pronotum	Black with anterior-lateral white marks, broadest at base	White with distinctive undulating-edged black pattern; undulations often fuse to leave white spots

▲ 11-spot Ladybird.

 life size

adult

larva

pupa

forms

Range
Notably coastal in northern and western parts of Britain and Ireland, with more inland records occurring in southern and central England.

National conservation status
Widespread.

Distribution trend (1995–2015)
Decreasing.

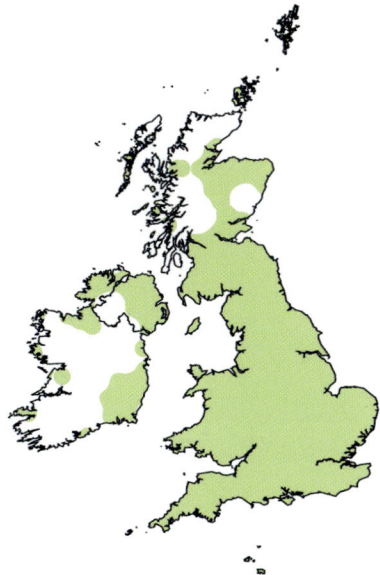

Harlequin Ladybird

Harmonia axyridis
(Pallas, 1773)

Large, domed and highly polymorphic ladybird that is native to central and eastern Asia and is sometimes also known as the 'multicoloured Asian ladybird'. The name 'Harlequin' stems from the nominate *axyridis* colour form of the species (a form very rarely recorded in Britain and Ireland), which resembles a characteristic harlequin chequered pattern.

Identification
Adult
Length 5–8mm
Background colour (1) f. *succinea*: yellow/orange/red; (2) f. *spectabilis* and (3) f. *conspicua*: black

Pattern (1) With 0–21 black spots; (2) with four red/orange spots/patches; (3) with two red/orange spots/patches
Number of spots 0–21 (16)
Spot fusions Common in *succinea* form
Other colour forms Common (*succinea*, *spectabilis* and *conspicua*)
Pronotum White or cream with up to five spots, or fused lateral spots forming two curved lines, M-shaped mark or solid trapezoid
Leg colour Brown
Other features Many specimens have a slight keel just before the tip of the elytra; extremely variable in colour and pattern

Confusion species				
	Harlequin Ladybird f. *succinea*	10-spot Ladybird f. *decempunctata*	Cream-streaked Ladybird ('16-spotted' form)	Eyed Ladybird
Size	5–8mm	3.5–4.5mm	5–6mm	7–8.5mm
Spots	0–21 black spots; usually two 'shoulder' spots at front of each elytron	0–15 black spots; usually one 'shoulder' spot at front of each elytron	16 black spots in a 1–3–3–1 pattern on each elytron	0–23 black spots, with or without cream rings around them
Pronotum	White with black M-mark or solid trapezoid marking	White with black spots	White with black spots	Black M-mark and distinctive black edging
Leg colour	Brown	Brown	Brown	Black

▲ Harlequin Ladybird f. *succinea*.

life size

larva

adult
f. *succinea*

pupa

f. *succinea*

f. *succinea*

f. *succinea*

f. *spectabilis*

f. *conspicua*

Late-instar larva Black, with thick dorsal spines coming from each tubercle, each branching at the top into three prongs. There is a bright orange upside-down L-shaped marking on each side, formed from the middle tubercles of abdominal segments one to five and the inner tubercles of abdominal segment one. Two pairs of orange dots on the dorsal surface, formed from the inner tubercles of abdominal segments four and five.

Pupa Orange, with pairs of black squarish markings running down the second thoracic segment and abdominal segments two to six; black and white remains of shed spiky larval skin visible at base of pupa.

Confusion species
Harlequin Ladybirds are highly variable. The most common form (f. *succinea*) is orange-red and may be confused with various species, particularly 10-spot and Cream-streaked Ladybirds. The 10-spot Ladybird is a markedly smaller species than the Harlequin Ladybird. The Cream-streaked Ladybird has streaky cream markings on its elytra and a more flattened shape.

Food
The Harlequin Ladybird feeds on aphids as well as many other insects, soft fruits, pollen and nectar.

Habitat
The Harlequin Ladybird is a habitat generalist and is considered to be arboreal, with many records from urban areas where deciduous trees are abundant. It is readily found on mature Lime or Sycamore trees, for example in churchyards and parks, but also occupies mature woodland (both deciduous and coniferous), scrub, grassland, marshland and reedbeds. Crops and orchards are other common habitats. It shows a preference for Lime and Sycamore trees but is also commonly associated with herbaceous plants such as Nettle, thistles, Cow Parsley, Rosebay Willowherb and Fat-hen, as well as ornamental plants in gardens. The Harlequin Ladybird overlaps with a number of other species of ladybird and is commonly found with 2-spot, 10-spot, 7-spot, 14-spot and Pine Ladybirds.

It is sometimes found in exceptionally large numbers in buildings during winter. Churches appear to be favoured, as well as domestic dwellings. Sheds, compost bins and, indeed, all sheltered locations in urban areas are suitable for overwintering.

Suggested survey method
Tree beating or visual searching of deciduous trees. In winter it can often be found indoors, for example around window-frames, ceilings and attics of houses. It is attracted to light and is commonly found in moth traps.

Range
Arrived in 2003, the first records in Britain coming from southeast England. Has since spread at a remarkable rate and is now widespread across England and Wales. Restricted distribution in Scotland, though recorded as far north as Orkney and Shetland. Has also been recorded in Ireland, the Channel Islands, the Isles of Scilly and the Isle of Man.

National conservation status
Ubiquitous.

Distribution trend (1995–2015)
Increasing.

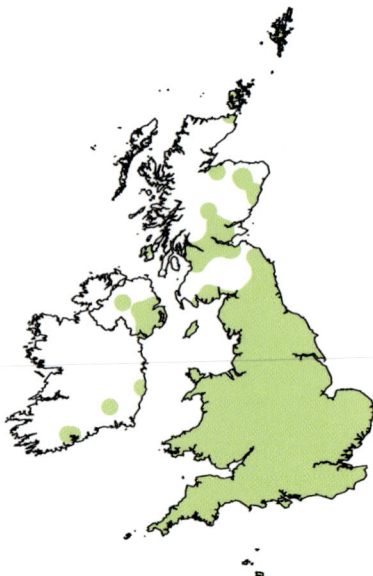

Cream-streaked Ladybird *Harmonia quadripunctata*
(Pontoppidan, 1763)

Rather a flat ladybird, with cream-streaked markings and two main forms, generally either having four or 16 spots, and only rarely any other number. A close relative of the Harlequin Ladybird.

Identification
Adult
Length 5–6mm
Background colour Pink, salmon, yellow
Pattern Black spots and cream streaking in two forms: (1) '16-spotted': 16 black spots in a 1–3–3–1 pattern on each elytron (most common) and (2) '4-spotted': 4 black spots on outer sides of elytra
Number of spots 4–20 (16)
Spot fusions Uncommon
Other colour forms Melanic (black) forms – rare
Pronotum White with 5–9 black spots in a distinctive pattern
Leg colour Brown
Other features Often rests head-down on pine buds, where it is very well camouflaged

Late-instar larva Black, with thick dorsal spines coming from each tubercle, each branching from the base into three prongs. There is a bright orange line on each side, formed from orange spots on the middle tubercles of abdominal segments one to four, and one pair of orange dots on the dorsal surface, formed from the inner tubercles of abdominal segment four.

Pupa Light greyish-brown, sometimes with a pink tinge, with six longitudinal rows of black spots and black lateral transverse markings on the anterior end; remains of shed spiky larval skin visible at the base of pupa.

Food
The Cream-streaked Ladybird is a predatory ladybird that feeds on aphids.

Habitat
The Cream-streaked Ladybird is a conifer specialist, but records have been received

109

▲ Cream-streaked Ladybird ('16-spotted' form).

adult
('16-spotted' form)

life size

larva

forms

pupa

from heathland, scrub, grassland and dune systems. It is the most common large ladybird found on conifers – usually Scots Pine but there are also records from exotic pines, Douglas Fir and Norway Spruce. It is occasionally found on herbaceous plants and shrubs such as Nettle and Gorse, but these are usually situated close to conifers.

The Cream-streaked Ladybird overwinters on various conifers, usually needled conifers but occasionally scale-leaved conifers such as Leyland Cypress.

Suggested survey method
Tree beating or visual searching of Scots Pine trees.

▲ Cream-streaked Ladybird egg-laying.

▲ Cream-streaked Ladybird ('4-spotted' form).

▲ Cream-streaked Ladybird ('16-spotted' form).

Range
Relatively new species to Britain, first recorded in 1937 in West Suffolk. Introduction pathway to Britain is unknown but is generally assumed to be natural spread from continental Europe, although human-mediated introduction is also plausible. The spread westwards and northwards across England is clear but slow; it took 50 years to reach as far west as Devon. It was first recorded in Scotland in 1982 (Berwickshire), with a few subsequent records from Scotland. Recorded from Ireland (Dublin coast) in 2017.

National conservation status
Local.

Distribution trend (1995–2015)
Stable.

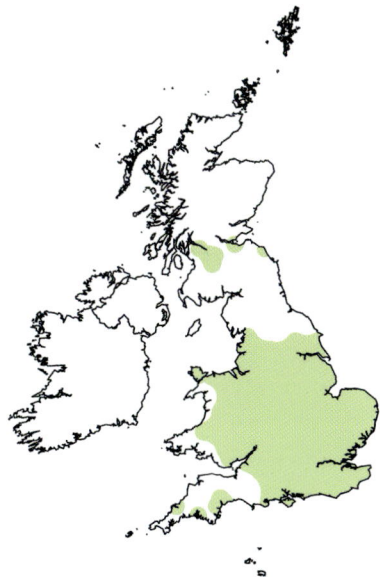

14-spot Ladybird   *Propylea quattuordecimpunctata*

(Linnaeus, 1758)

Medium-sized yellow ladybird with distinctive rectangular black spots that are often fused and resemble an anchor shape or clown's face. Sometimes referred to as the 'dormouse' ladybird because it emerges from overwintering later than many of the other conspicuous species.

112

Identification

Adult
Length 3.5–4.5mm
Background colour Yellow or rarely black
Pattern Black or rarely yellow spots
Number of spots 4–14 (14)
Spot fusions Very common
Other colour forms Melanic (black) forms – very rare
Pronotum Yellow or cream with black spots or trapezium or crown-shaped mark
Leg colour Brown
Other features Spots are rather rectangular

Late-instar larva Highly characteristic, with long legs and distinctive markings. Dark greyish-brown with cream patches surrounding tubercles on thoracic segments; all outer tubercles, the middle tubercles of the first abdominal segment and all tubercles of the fourth abdominal segment are cream. Cream markings along the midline.

Pupa Pale brown, but with slightly darker thoracic segments; third thoracic segment bearing a pair of dark spots flanking the midline; a band of yellow to cream spots across the first thoracic segment. Four rows of diffuse dark markings running longitudinally down abdominal segments two to eight.

Confusion species

The 14-spot Ladybird is sometimes confused with the 10-spot Ladybird, particularly the 'chequered' form (f. *decempustulata*) of the latter. They occur in the same habitats. The best distinguishing feature is the pronotum, but also the spot patterning.

Food

Aphids; considered a major aphid predator in agricultural systems.

Habitat

The 14-spot Ladybird occupies many diverse habitats such as grasslands, saltmarsh and scrub. It also occurs in mature woodlands (both deciduous and coniferous) and orchards, and is commonly found on crops, particularly broad-leaved crops such as Field Bean but also cereals. Most common on low herbaceous vegetation, particularly Nettle but also Foxglove, Angelica, docks, Cow Parsley, Hogweed, Mugwort, Tansy, thistles, Rosebay Willowherb and Fat-hen. Cultivated beans, Blackcurrant and roses are popular

▲ 14-spot Ladybird.

Confusion species		
	14-spot Ladybird	10-spot Ladybird f. *decempustulata*
Size	3.5–4.5mm	3.5–4.5mm
Spots	Usually 14 often fused	Often 10 in grid-like markings giving a chequered pattern
Pronotum	Yellow or cream, usually with crown-shaped mark	White with five dark spots, which may be fused

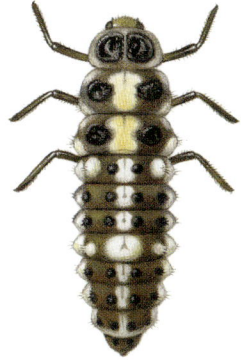

life size

113

larva

adult

forms

pupa

host plants for 14-spot Ladybirds in gardens. Also, there are many records of this species from deciduous trees and shrubs including Hawthorn, sallow, willow, Sycamore, Lime, oak and birch. Overlaps with a number of other species of ladybird including 2-spot, 10-spot, 7-spot, Pine and Harlequin Ladybirds.

The 14-spot Ladybird overwinters in many habitats, commonly on low herbage.

Suggested survey method
Sweep netting or visual searching of low herbaceous plants.

Range
One of the commonest species of ladybird in England, Wales and Ireland, but there are few Scottish records.

National conservation status
Ubiquitous.

Distribution trend (1995–2015)
Stable.

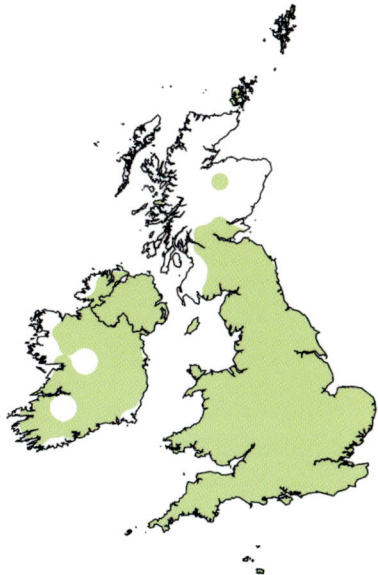

Eyed Ladybird

Anatis ocellata
(Linnaeus, 1758)

The largest ladybird in Britain. Often with pale rings around the black spots on its elytra (although some specimens lack the rings; even more rarely, the black centres are absent).

114

Identification
Adult
Length 7–8.5mm
Background colour Russet or burgundy
Pattern Black spots, with or without cream rings around them (sometimes spots absent)
Number of spots 0–23 (15)
Spot fusions Rare
Other colour forms Melanic (black) forms – very rare
Pronotum Distinctively patterned – white markings with black M-mark
Leg colour Black
Other features Largest ladybird in Britain and Ireland

Late-instar larva Large (can exceed 12mm), with thick black spines projecting from tubercles. There are two large orange spots on the outer tubercles of the first two abdominal segments, and pairs of small white or yellowish spots along the bottom edges of abdominal segments five to eight.

Pupa Cream with four rows of black spots running longitudinally and an additional pair of black spots on the sides of abdominal segments two and three. There is a black zigzag across the thoracic segments, and three ragged teeth on each side of the middle section give a sculptured appearance, spanning segments three to five.

Food
The Eyed Ladybird is a predatory ladybird that feeds on aphids.

Habitat
The Eyed Ladybird is a conifer specialist and is most often found in coniferous and mixed woodlands. The majority of Eyed Ladybird records are associated with conifers, particularly Scots Pine, but also Douglas Fir and larches. In late summer this species may be found on deciduous trees such as Lime and oak. There are occasional records from Nettle and other herbaceous plants, but usually in the vicinity of conifers.

▲ Eyed Ladybird.

life size

adult

larva

forms

pupa

The Eyed Ladybird overwinters in the soil or leaf litter in coniferous or mixed woodlands.

Suggested survey method
Tree beating or visual searching of Scots Pine trees.

Range
Very widespread species in Britain and Ireland, and has been recorded as far north as Shetland.

National conservation status
Widespread.

Distribution trend (1995–2015)
Decreasing.

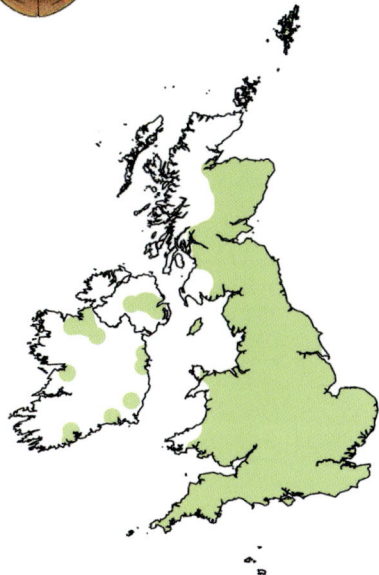

18-spot Ladybird

Myrrha octodecimguttata
(Linnaeus, 1758)

Maroon-brown ladybird with cream spots which vary in shape and are sometimes fused; two spots at the front of the elytra are distinctively L-shaped.

Identification
Adult
Length 4–5mm
Background colour Maroon-brown
Pattern Cream spots
Number of spots 14–18 (18)
Spot fusions Common
Other colour forms None
Pronotum Creamy-white with rounded M-mark
Leg colour Brown

Late-instar larva Quite smooth, pale grey body with pairs of stippled dark grey patches on thoracic segments; middle and outer tubercles on first abdominal segment pale yellow; all other tubercles dark grey with a few short hairs protruding. Legs bicoloured dark and pale. Very similar to larva of Larch Ladybird (but note leg colour difference), so rearing through to adulthood is recommended for accurate identification.

Pupa Cream, with two rows of diffuse brown markings flanking the midline; similar markings found laterally along abdominal segments three to eight; yellow lateral markings on first abdominal segment.

Confusion species
The 18-spot Ladybird is commonly confused with the Cream-spot Ladybird, which is a species commonly associated with deciduous trees and so is generally not found in the same habitat. Colour of the two species is similar, a maroon-brown with creamy-white spots, but pattern of spotting is different, Cream-spot Ladybirds having 14 round spots on the elytra, while the spots of 18-spot Ladybirds are variable in shape and sometimes fused; two spots at the front of the elytra are distinctively L-shaped in 18-spot Ladybirds.

Food
The 18-spot Ladybird is a predatory ladybird that feeds on aphids.

Habitat
The 18-spot Ladybird is a conifer specialist, occurring mainly in coniferous woodlands but also on scrub, heathland, grassland and dune systems where conifers are present. It is most commonly recorded on Scots Pine but other conifers are also occupied, including Black Pine and Monterey Cypress. There are

Confusion species		
	18-spot Ladybird	Cream-spot Ladybird
Size	4–5mm	4–5mm
Spots	18 variably shaped and sometimes fused cream spots; two spots at front of elytra distinctively L-shaped	14 distinct round cream spots
Pronotum	Creamy-white with rounded maroon M-marking	Maroon with cream markings at the side
Habitat	Conifer specialist	Many deciduous trees

▲ 18-spot Ladybird.

116

life size

adult

larva

pupa

form

occasional records from deciduous trees, including oak and Sycamore, but these are unlikely to be favoured for reproduction.

The 18-spot Ladybird overwinters high in the crown of Scots Pine trees, and under the bark.

Suggested survey method
Tree beating or visual searching of Scots Pine trees.

Range
Locally distributed throughout England and Wales. Limited distribution in Scotland and Ireland.

National conservation status
Local.

Distribution trend (1995–2015)
Stable.

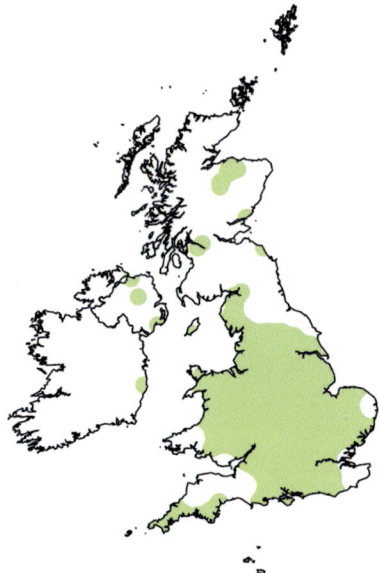

Cream-spot Ladybird   *Calvia quattuordecimguttata*
(Linnaeus, 1758)

Maroon-brown ladybird with distinctive round cream spots. Female Cream-spot Ladybirds deposit a red waxy substance on the surface of their eggs, which is visible with a hand lens.

118

Identification
Adult
Length 4–5mm
Background colour Maroon-brown
Pattern Cream spots
Number of spots 14
Spot fusions Very rare
Other colour forms Melanic (black) forms – very rare
Pronotum Maroon with cream marks at the sides

▲ Cream-spot Ladybird.

Leg colour Brown
Other features Six of the spots form a line across the elytra

Late-instar larva Dark grey, with two pairs of bold white marks on the middle and outer tubercles of the first and fourth abdominal segments. The outer white tubercles of abdominal segments four to six are tall and pointed, giving a saw-toothed appearance to the side of the body.

Pupa Creamy-yellow with diffuse dark grey/brown markings running longitudinally. There are three ragged teeth on each side of middle section, spanning segments three to five.

Confusion species
The Cream-spot Ladybird is commonly confused with the Orange Ladybird, although the latter usually appears as bright orange and the former maroon-brown. It is also similar to the 18-spot Ladybird.

Food
The Cream-spot Ladybird is a predatory ladybird that feeds on aphids and psyllids.

Habitat
The Cream-spot Ladybird is commonly found in woodland, but there are a number of records from grassland, heathland and marshland. It is a deciduous tree specialist and lives among

Confusion species			
	Cream-spot Ladybird	Orange Ladybird	18-spot Ladybird
Size	4–5mm	4.5–6mm	4–5mm
Background colour	Maroon-brown	Orange	Maroon-brown
Spots	14 distinct round cream spots often appearing in ordered transverse rows (1–3–2–1 from front to rear of each elytron)	16 white spots appearing more scattered across the elytra than in the Cream-spot Ladybird	18 variably shaped and sometimes fused cream spots; two spots at front of elytra distinctively L-shaped
Pronotum	Maroon with lateral cream markings	Orange with translucent edging	Creamy-white with rounded maroon M-marking
Leg colour	Brown	Orange	Brown
Habitat	Many deciduous trees	Many deciduous trees	Conifer specialist

life size

adult

larva

pupa

the foliage of broad-leaved deciduous trees, hedges and shrubs. The species is particularly associated with Ash but is also commonly found on Sycamore, Lime, Beech and oak. There are also a number of records from Hawthorn, Hornbeam, sallow and Gorse. Records on conifers are scarce, but include some from Scots Pine. Some records are from herbaceous plants such as thistles and Nettle.

The Cream-spot Ladybird overwinters close to the ground in the leaf-litter layer, bark crevices and Beech nuts.

Suggested survey method
Tree beating or visual searching of deciduous trees.

Range
Widely distributed across Britain and Ireland.

National conservation status
Ubiquitous.

Distribution trend (1995–2015)
Stable.

Striped Ladybird

Myzia oblongoguttata
(Linnaeus, 1758)

The most specialised of all the aphid-feeding ladybirds in Britain; largely restricted to mature Scots Pine trees. Second-largest ladybird in Britain and Ireland; chestnut colour with pale stripy markings, making it very distinctive and difficult to confuse with any other species. Well camouflaged on Scots Pine trees. The pronotum is sometimes a darker brown than the elytra.

Identification
Adult
Length 6–8mm
Background colour Chestnut-brown
Pattern Cream stripes and spots
Number of spots 0–15 (13)
Spot fusions Common
Other colour forms Melanic (black) forms – rare
Pronotum White with chestnut M-mark or trapezium
Leg colour Brown

Late-instar larva Large (up to 12mm), with long black legs. The body has an overall smooth appearance (tubercles do not bear spines or hairs) and is grey with contrasting dark grey/black tubercles. Outer and middle tubercles of first abdominal segment are orange, as are the outer tubercles of the fourth and sixth segments.

Pupa Projections give sculptured appearance. Cream with four rows of black spots running longitudinally and an additional black spot on the outer edges of the third abdominal segment; third thoracic segment flanked by thick black diagonal markings; rounded projections on each side of middle section, spanning segments two to five.

Food
The Striped Ladybird is largely restricted to feeding on large brown aphids of the genus *Cinara*, which occur on pines.

Habitat
This species is a conifer specialist but has been reported on scrub and dune systems with conifers in the vicinity. Most commonly found on Scots Pine, but there are a few records from larch.

The Striped Ladybird overwinters in soil or moss below Scots Pine trees.

▲ Striped Ladybird.

life size

adult

larva

pupa

Suggested survey method
Tree beating or visual searching of Scots Pine trees.

Range
Locally distributed and generally less common and abundant than most of the other ladybirds found on pines, at least in England. Unlike several other ladybird species, it is tolerant of cool conditions and is widespread in pine forests across Scotland.

National conservation status
Very local.

Distribution trend (1995–2015)
Stable.

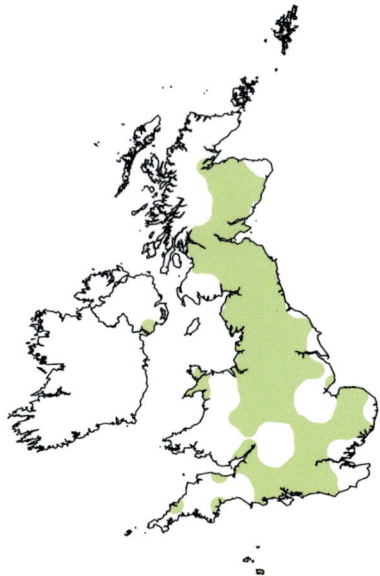

Bryony Ladybird

Henosepilachna argus
(Geoffroy in Fourcroy, 1762)

Large ladybird with slightly fuzzy appearance due to downy hairs on elytra.

Identification
Adult
Length 5–7mm
Background colour Orange
Pattern Black spots
Number of spots 11
Spot fusions Rare
Other colour forms None
Pronotum Orange
Leg colour Orange
Other features Elytra covered in short downy hairs

Late-instar larva Pale yellow with dark tubercles, bearing tall black branching spines.

Pupa Pale yellow with small black spots; partially covered by shed larval skin at base.

Food
The Bryony Ladybird feeds on members of the cucumber (cucurbit) family, White Bryony in particular.

Habitat
In Britain the Bryony Ladybird is found in urban habitats, commonly gardens, allotments and car parks. In Surrey there are recent sightings from more natural habitats on chalk and sand. In Britain it feeds solely on White Bryony but in other parts of Europe it has been noted feeding on Melon.
It overwinters in low herbage.

Suggested survey method
Visual searching of White Bryony.

Range
Recent colonist of Britain, having established in southwest London in the 1990s. In Europe, spread has been linked to recent changes in climate.

National conservation status
Very local.

Distribution trend (1995–2015)
Stable.

▲ Bryony Ladybird.

life size

adult

larva

pupa

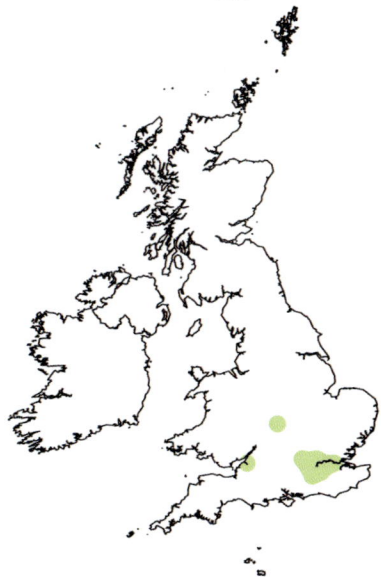

24-spot Ladybird
Subcoccinella vigintiquattuorpunctata (Linnaeus, 1758)

Small and hairy species; often missed when searching for ladybirds by eye.

Identification
Adult
Length 3–4mm
Background colour Russet
Pattern Black spots
Number of spots 0–24 (20)
Spot fusions Common
Other colour forms Melanic (black) forms – rare
Pronotum Russet with black spots
Leg colour Russet
Other features Elytra covered in fine hairs (visible with hand lens), giving the ladybird a matt appearance

Late-instar larva Cream-yellow or greenish in colour and short and stubby in shape. Tubercles are dark, bearing thick yellow spiny bristles with extensive side-branching.

Pupa Pale yellow with small black spots, partially covered by shed larval skin at base.

Food
The 24-spot Ladybird is a plant-feeding (phytophagous) species that generally feeds on the leaves of Red Campion in England, and on Lucerne in southeastern Europe, although Roger Hawkins' (2000) survey of the ladybirds of Surrey revealed extensive populations feeding on False Oat-grass. This is also a common host plant in East Anglia and probably elsewhere.

Habitat
The 24-spot Ladybird is a grassland species that favours 'rough grass'; however, records have also been received from marshy habitats and scrub. It is most commonly found on grasses but also occurs on other low-growing plants such as thistles, Nettle, Mugwort, Salad Burnet, knapweed, Spurrey and Tansy.

It overwinters in low herbage, grass tussocks and Gorse bushes.

Suggested survey method
Sweep netting of grassland.

Range
Widely distributed in England and Wales. Limited distribution in Scotland; absent from Ireland (but there is one old record).

National conservation status
Widespread.

Distribution trend (1995–2015)
Stable.

▲ 24-spot Ladybird.

life size

adult

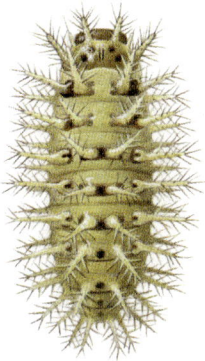

larva

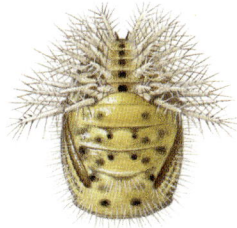

pupa

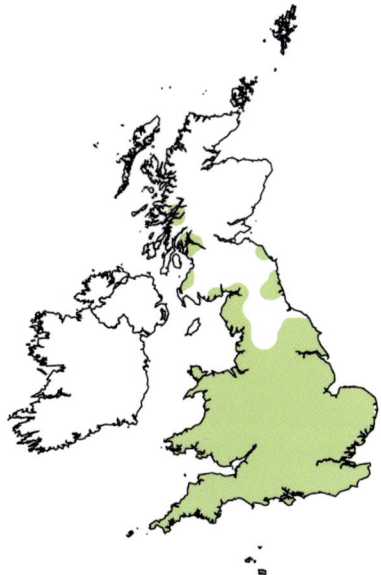

Red Marsh Ladybird

Coccidula rufa
(Herbst, 1783)

Coccidula rufa is the best recorded of the inconspicuous ladybirds. This widespread and common species is quite distinct from most ladybirds because it has an elongate body and relatively long antennae.

Identification (adult)
Length 2.5–3mm
Background colour Brownish-red
Pattern Ill-defined dark elytral marking either side of the scutellum
Number of spots 0
Pronotum Brownish-red
Leg colour Brownish-red
Other features Hairy; elongate (oblong with parallel sides) and flattened; long antennae; brownish-red head

Confusion species
Coccidula rufa is similar to *C. scutellata*, but is plain brownish-red with no spots (whereas *C. scutellata* has spots on the elytra).

Food
Coccidula rufa is a predatory ladybird that feeds on aphids.

Habitat
Coccidula rufa is usually found in wet habitats including wet grasslands, fens, marshes (and sometimes saltmarshes), mires, riversides, lake and pond edges, ditches and even dunes. The vast majority of British and Irish records come from these habitats, where it is generally found on plants such as Reed, rushes, Reedmace and wetland grasses. There are also a few woodland records of the species (especially in wet woodlands) on willow, Alder, birch and oak. It has rarely been recorded on plants such as Gorse and rose.

C. rufa overwinters in the leaf sheaths of Reed, rushes and Reedmace and in tufts of grass.

Suggested survey method
Sweep netting in Reed or wetland grasses; visual searching within Reed stems in winter. It is easiest to find when it is overwintering in the leaf sheaths of Reed or Reedmace.

Range
Widely distributed throughout Britain and Ireland.

National conservation status
Very widespread.

Distribution trend (1995–2015)
Decreasing.

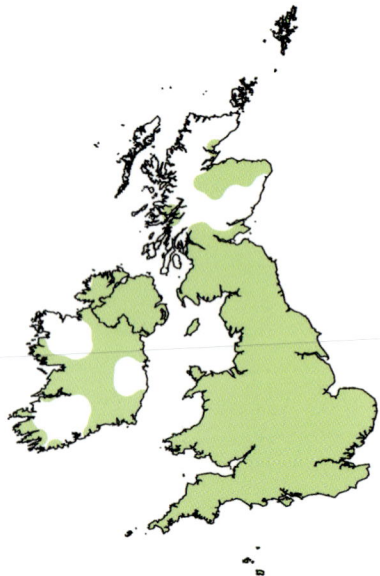

life size

adult

Spotted Marsh Ladybird

Coccidula scutellata
(Herbst, 1783)

Similar in appearance and size to the more common *Coccidula rufa*, but *C. scutellata* can be clearly distinguished by the presence of five black spots on the elytra.

Identification (adult)
Length 2.5–3mm
Background colour Brownish-red
Pattern Black spots
Number of spots 5
Pronotum Brownish-red, narrower than elytra
Leg colour Brownish-red
Other features Hairy; elongate (oblong with parallel sides) and dorsoventrally flattened; long antennae; head brownish-red

Food
Coccidula scutellata is a predatory ladybird that feeds on aphids.

Habitat
Coccidula scutellata is found in marshes, riversides and pondsides, often on Reedmace, Reed and rushes. It has also been recorded on other plants in wet habitats, including Yellow Iris. There is one unusual record of the species being found in the stomach of a Blue Tit.

 C. scutellata overwinters in the leaf sheaths of Reedmace and Reed.

Suggested survey method
Sweep netting in Reed or wetland grasses; visual searching within Reed stems in winter.

Range
Much more restricted distribution than *C. rufa*; although widespread in England (including close to the Scottish border), there are no records of *C. scutellata* in Scotland or Ireland.

National conservation status
Very local.

Distribution trend (1995–2015)
Decreasing.

127

adult

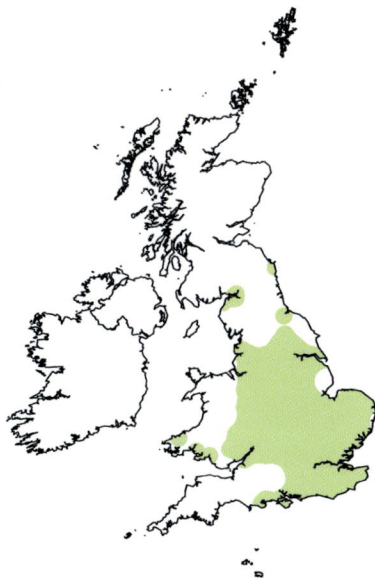
life size

Round-keeled Rhyzobius   *Rhyzobius chrysomeloides*
(Herbst, 1792)

Small ladybird with a dorsal covering of short downy hairs.

Identification (adult)
Length 2.5–3.5mm
Background colour Pale to dark brown
Pattern Variable, but often with dark U-shaped mark towards rear of elytra
Number of spots 0
Pronotum Pale to dark brown
Leg colour Pale to dark brown
Other features Hairy; long antennae; head pale to dark brown

Confusion species		
	R. chrysomeloides	*R. litura*
Elytra	Often paler, with more extensive dark markings	Pale brown, with less extensive dark markings
Prosternal keel	Broader, with rounded apex	Triangular, with pointed apex
Habitat	Trees	Low-growing vegetation

Food
Rhyzobius chrysomeloides is a predatory ladybird that feeds on aphids.

Habitat
Rhyzobius chrysomeloides may be found on a range of trees in various habitats, including gardens. It favours pine, but has been recorded on other conifers in Britain, including cypress, Juniper, larch and fir. There are some records from deciduous trees, such as oak, willow, maple and Hawthorn. It also seems to favour Ivy, and has been recorded on various garden shrubs including viburnum, Firethorn and oleaster.

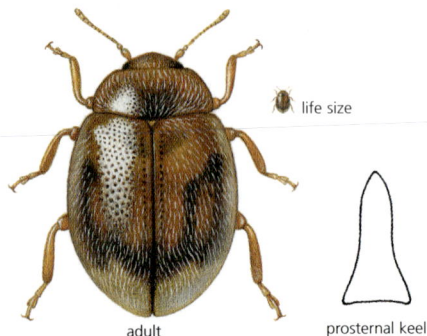

Overwintering sites in Britain are unclear, but there are a few winter records from the evergreen garden shrubs mentioned above.

Suggested survey method
Tree beating.

Range
First found in Britain in 1996 on a pine tree on a Surrey motorway bank, and well established in parts of Surrey by the late 1990s. It has since spread across southeast England and in recent years has been found in new counties such as Oxfordshire, Worcestershire, Cambridgeshire and Northamptonshire. Some records are from urban habitats.

National conservation status
Very local.

Distribution trend (1995–2015)
Increasing.

life size

adult

prosternal keel

Pointed-keeled Rhyzobius

Rhyzobius litura
(Fabricius, 1787)

Small ladybird with a dorsal covering of short downy hairs. Varies in colour from pale yellow to dark brown, often with a dark squarish U-shaped mark on the elytra. Larvae and pupae are lemon-yellow and covered in long fine hairs. The adult's cryptic colouration makes it difficult to spot and it is often overlooked, even by ladybird enthusiasts.

Identification (adult)
Length 2.5–3mm
Background colour Pale to dark brown
Pattern Variable, but often with dark U-shaped mark towards rear of elytra
Number of spots 0
Pronotum Pale to dark brown
Leg colour Pale to dark brown
Other features Hairy; long antennae; head pale to dark brown

Confusion species
R. chrysomeloides; see opposite.

Food
Rhyzobius litura is a predatory ladybird that feeds on aphids.

Habitat
Rhyzobius litura is a grassland species that may be found on low-growing vegetation, especially grasses, thistles and Nettle. It has often been recorded in dune systems and has sometimes been found in wet habitats, on rushes and

Reed. There are a few records from shrubs, including Gorse, and trees, including willow.
   Overwintering sites tend to be in low-growing vegetation, especially grass tussocks and sometimes mosses.

Suggested survey method
Sweep netting in grassland or Nettle beds.

Range
Well recorded in comparison to many of the other inconspicuous ladybirds, and in many areas can easily be found when sweep netting in meadows or grass verges. Probably one of the more common ladybird species, at least in England and Wales. Most of the records of *R. litura* in Wales, Scotland and Ireland are coastal, but in England there are many inland records.

National conservation status
Widespread.

Distribution trend (1995–2015)
Stable.

129

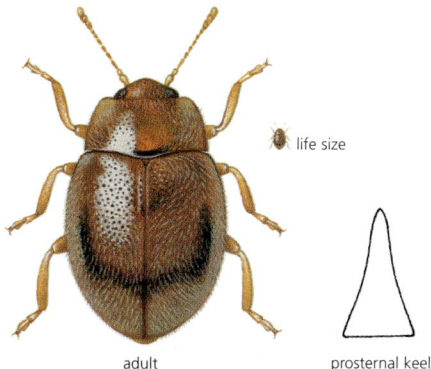
adult   prosternal keel

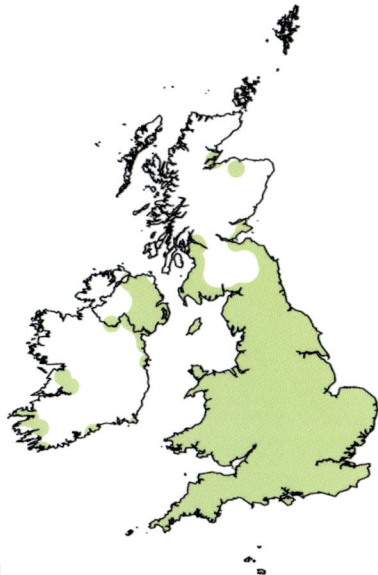
life size

Red-headed Rhyzobius

Rhyzobius lophanthae
(Blaisdell, 1892)

Small black ladybird with a dull orange pronotum and head. *Rhyzobius lophanthae* is smaller and has shorter antennae than the other *Rhyzobius* species found in Britain, and has different and distinctive colouration.

Identification (adult)
Length 2mm
Background colour Black
Pattern None
Number of spots 0
Pronotum Dull orange
Leg colour Dull orange
Other features Entire dorsal surface covered in short hairs; long antennae; head dull orange; margins of the elytra with long hairs

Food
Rhyzobius lophanthae is a predatory ladybird that feeds on scale insects. It was widely used in Europe throughout the twentieth century as a biological control agent of armoured scale insects and is now widespread around the Mediterranean basin.

Habitat
Rhyzobius lophanthae can be quite urban, and many of the records are from parks and gardens. It is often found on cypress trees, particularly Leyland Cypress, with other records from Juniper, viburnum and spindle. However, it has more rarely also been recorded on broad-leaved trees such as oak, Ash and Lime.

Overwintering sites are largely unknown, but are probably similar to its breeding habitat, for example evergreen trees such as cypresses.

Suggested survey method
Tree beating.

Range
Rhyzobius lophanthae is native to Australia. First found in Britain by D.A. Coleman on an Ash tree in Morden Park, Surrey, in 1999. It has since been recorded as breeding outdoors in London. It has been spreading quickly over the last 10 years, and there are recent records from many counties across southern England and as far north as Nottinghamshire and Lincolnshire. There are records from all times of year, so the species is evidently now surviving winters in Britain (something that seemed in doubt when it was first recorded).

National conservation status
Very local.

Distribution trend (1995–2015)
Insufficient data.

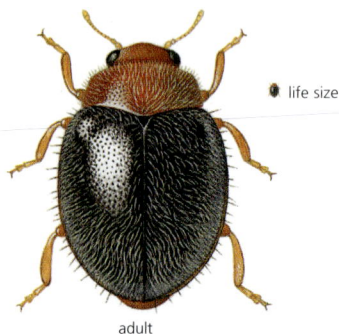

life size

adult

False-spotted Ladybird *Hyperaspis pseudopustulata*
Mulsant, 1853

Unlike other inconspicuous ladybirds, *Hyperaspis pseudopustulata* is shiny, hairless and reasonably large, and is easily recognisable as a ladybird. It is black with two small orangey-red spots towards the tips of the elytra. However, it is not an easy insect to find.

Identification (adult)
Length 3–4mm
Background colour Black
Pattern Small orangey-red spots near tips of elytra
Number of spots 2
Pronotum Black with orangey-red lateral and, often, anterior margins
Leg colour Black and brown
Other features Hairless; mouthparts and antennae brown; head black and red

Food
Hyperaspis pseudopustulata is a predatory ladybird that feeds on scale insects and aphids.

Habitat
Hyperaspis pseudopustulata has diverse habitat requirements but is often coastal or found in wet habitats. It favours low vegetation at water margins, including Reed and Reedmace.

    H. pseudopustulata overwinters in leaf litter and moss.

Suggested survey method
Sweep netting.

Range
While *H. pseudopustulata* is widespread in England and Wales, there are few recent records of it there, and no recent records from Scotland or Ireland.

National conservation status
Very local.

Distribution trend (1995–2015)
Insufficient data.

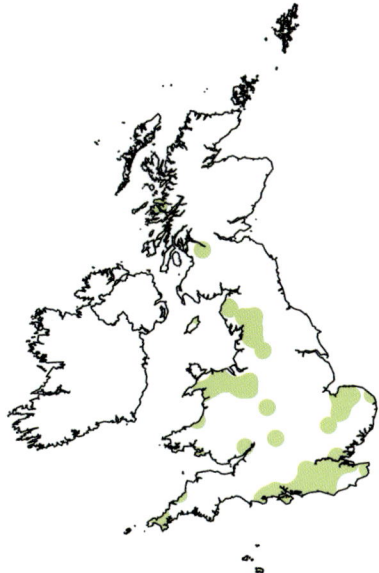

life size

adult

Horseshoe Ladybird

Clitostethus arcuatus
(Rossi, 1794)

Tiny ladybird with an attractive and distinctive horseshoe-shaped mark on its elytra (easily visible when viewed through a hand lens).

Identification (adult)
Length 1.2–1.5mm
Background colour Dark brown to black
Pattern Pale yellow/cream horseshoe-shaped mark in the centre of the elytra
Number of spots 0
Pronotum Dark brown but cream at sides, or sometimes mostly cream
Leg colour Cream
Other features Hairy; head dark brown to black, or sometimes cream

Food
Clitostethus arcuatus is a predatory ladybird that feeds on whitefly.

Habitat
Clitostethus arcuatus is found in coniferous and deciduous woodland or other habitats with trees. It favours Ivy, but has also been recorded on Honeysuckle, viburnum and Holly.

   C. arcuatus overwinters in bark crevices and under the bark of coniferous and deciduous trees.

Suggested survey method
Beating appropriate shrubs. Most often found by beating Ivy growing on trees such as oak.

Range
The very few records partly reflect this species' preference for warm climates; it is more common in Mediterranean countries. Indeed, most of the few records that we do have are southerly and are recent, including Worcestershire records in 2011 and 2017. It may be that the species is becoming more abundant with climate warming. It is probably also under-recorded, partly because of its very small size.

National conservation status
Very local.

Distribution trend (1995–2015)
Insufficient data.

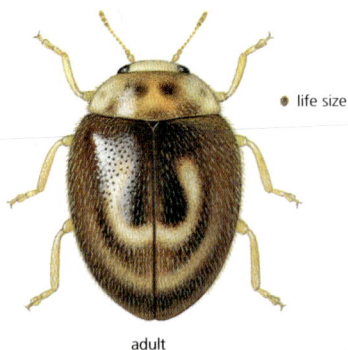

132

● life size

adult

Two-spotted Nephus

Nephus bisignatus
(Boheman, 1850)

This small species often has two reddish-brown spots towards the base of the elytra, but these are variable and sometimes absent. Still on the official British list but is now thought to be extinct in Britain (Majerus 1994), although there is a single record, from Rye Harbour, East Sussex, in May 1996.

Identification (adult)
Length 1.5–2mm
Background colour Black
Pattern Variable, but often with two reddish-brown spots towards rear of elytra
Number of spots 2
Pronotum Black
Leg colour Brown
Other features Fine reddish-brown margin to base of elytra; head black

Food
Nephus bisignatus is a predatory ladybird that feeds on scale insects. Can be important in controlling mealybug populations, for example in vineyards.

Habitat
Nephus bisignatus is a species of woodlands and other habitats with trees. It possibly favours Leyland Cypress and other related species.
   The overwintering sites in Britain are unknown.

Suggested survey method
Tree beating.

Range
Other than the single record in 1996, there are two known localities for the species (Deal, Kent, and Pevensey Bay, East Sussex), both from pre-1940 museum records. In Europe, *N. bisignatus* has a wide distribution, including Scandinavia, the Netherlands, Germany, southern France, Italy, Portugal and Greece.

National conservation status
Extinct.

Distribution trend (1995–2015)
Insufficient data.

133

adult

life size

Four-spotted Nephus

Nephus quadrimaculatus
(Herbst, 1783)

Bears four distinctive markings on its elytra and can be commonly seen on Ivy covering south-facing walls.

Identification (adult)
Length 1.5–2mm
Background colour Black
Pattern Two pairs of reddish-brown kidney-shaped spots, the front pair being larger than the rear pair
Number of spots 4
Pronotum Black
Leg colour Brown
Other features Hairy; abdomen has a reddish-brown tip; head black

Food
Nephus quadrimaculatus is a predatory ladybird that feeds on scale insects.

Habitat
Nephus quadrimaculatus may be found in gardens, woodlands (deciduous and coniferous) and other habitats where Ivy is prevalent. It can be found (sometimes in large numbers) in patches of Ivy on garden walls and trees, as well as in woodlands. It has also been recorded on Alder and Sycamore (not always with Ivy) and shrubs such as Firethorn.

  N. quadrimaculatus probably overwinters primarily in Ivy.

Suggested survey method
Beating or visual searching of Ivy.

Range
Once considered a rare species, with most records coming from Suffolk, it was found during the 1990s in other southeastern counties, particularly Surrey, where it is now common. It is still largely restricted to southeast England (but at least as far north as Norfolk and Cambridgeshire), and most of the records have been made since 2000. However, there are various recent records from more westerly areas, including Worcestershire and Shropshire, suggesting that the species may be increasing and spreading.

National conservation status
Very local.

Distribution trend (1995–2015)
Stable.

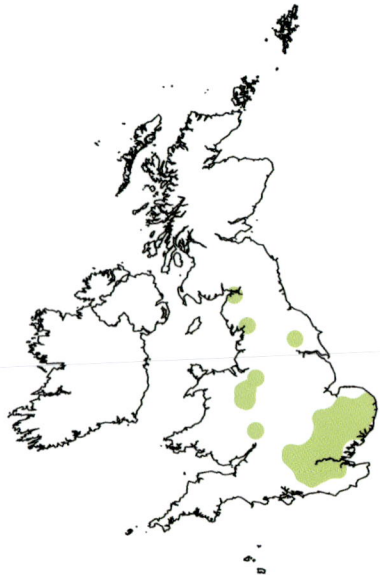

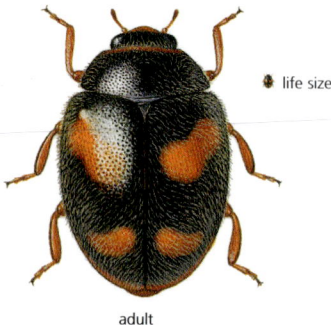

🐞 life size

adult

Red-patched Nephus

Nephus redtenbacheri
(Mulsant, 1846)

Black elytra of *Nephus redtenbacheri* bear characteristic elongate reddish-brown patches.

Identification (adult)
Length 1.3–2.3mm
Background colour Black
Pattern Two large elongate reddish-brown ill-defined patches, which are variable in shape
Number of spots 2
Pronotum Black
Leg colour Light to dark brown
Other features Hairy; head black

Confusion species
Can be confused with *Scymnus limbatus* but it is smaller and more elongate. Additionally, it is found in different habitats to *S. limbatus*, which favours deciduous trees.

Food
Nephus redtenbacheri is a predatory ladybird that feeds on mealybugs (a type of mobile scale insect) and aphids.

Habitat
Nephus redtenbacheri may be found in undisturbed grassland, dunes, heathland, bogs and even deciduous woodland. Tends to favour coastal habitats, including dunes, where it is found in low-growing vegetation. There are also many inland records.

N. redtenbacheri overwinters in low herbage, leaf litter and moss.

Suggested survey method
Sweep netting.

Range
A habitat generalist that is widespread across Britain and Ireland.

National conservation status
Very local.

Distribution trend (1995–2015)
Decreasing.

135

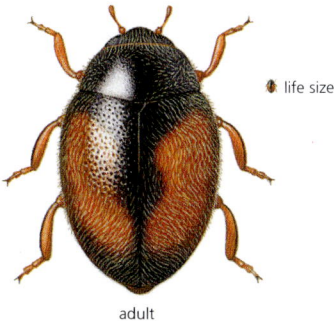

◗ life size

adult

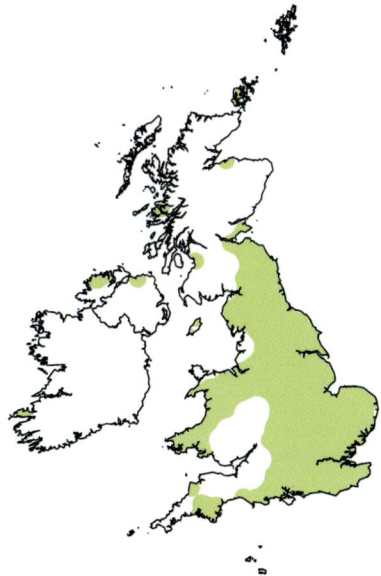

Red-rumped Scymnus

Scymnus haemorrhoidalis
Herbst, 1797

Distinctive broad red tip to the elytra, and the head and pronotum are also mostly red. Supposedly a common species, but difficult to find because it lives close to the ground, rather than higher up on low vegetation; hence it may be under-recorded.

Identification (adult)
Length 1.8–2.2mm
Background colour Black
Pattern Broad reddish-brown patch at rear of elytra
Number of spots 0
Pronotum Black with reddish-brown front margin
Leg colour Brown
Other features Hairy; tip of abdomen is reddish-brown; head reddish

Food
Scymnus haemorrhoidalis is a predatory ladybird that feeds on aphids.

Habitat
Scymnus haemorrhoidalis is a species of damp habitats including bogs, water margins and undisturbed grassland. It lives low down on low-growing vegetation, such as grass tussocks and moss. It may occasionally be found on shrubs and trees, Reed, leaf litter or the rosettes of perennial plants.

S. haemorrhoidalis overwinters low down in low herbage.

Suggested survey method
Sweep netting or visual searching of appropriate vegetation close to ground level.

Range
Widespread species in southern Britain and Jersey, but there are no records from Scotland or Ireland. Records of the species are sparse, especially in recent years.

National conservation status
Very local.

Distribution trend (1995–2015)
Decreasing.

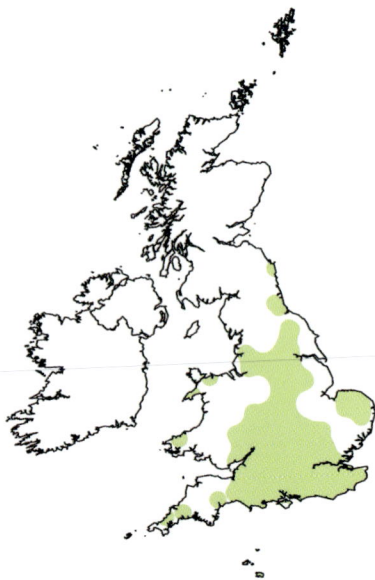

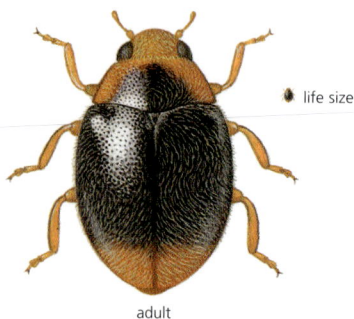

● life size

adult

Bordered Scymnus

Scymnus limbatus
Stephens, 1832

Hairy, brown with dark strip running along midline where elytra meet. Similar in appearance to *Nephus redtenbacheri* but is larger and found in deciduous trees in marshy habitats. *Scymnus limbatus* is an uncommon species, and there are very few recent records of it.

Identification (adult)
Length 1.6–2mm
Background colour Brown
Pattern A broad dark brown to black strip running along the centre line, widening at the top and bottom
Number of spots 0
Pronotum Dark brown to black
Leg colour Brown, although variable
Other features Hairy; head dark brown to black

Food
Scymnus limbatus is a predatory ladybird that feeds on aphids and scale insects.

Habitat
Scymnus limbatus is a species of deciduous trees in marshy habitats. Host plants are mainly willows and poplars, though it is sometimes found on other deciduous trees, such as elm.

It overwinters in bark crevices and under the bark of willows and poplars.

Suggested survey method
Tree beating.

Range
Most records are from England, the exception being an early twentieth-century record from Summerhill, County Fermanagh, Northern Ireland.

National conservation status
Very local.

Distribution trend (1995–2015)
Insufficient data.

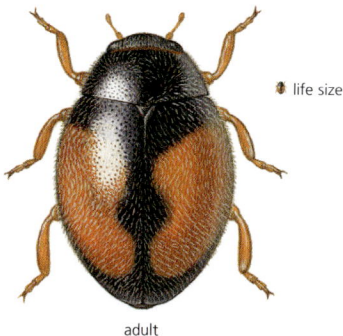

life size

adult

Oak Scymnus

Scymnus auritus
Thunberg, 1795

Slightly larger than many of our *Scymnus* species, *S. auritus* is a highly specialised feeder, eating aphid-like insects of the family Phylloxeridae, and is usually found on oak trees.

138

Identification (adult)
Length 2–2.3mm
Background colour Black
Pattern Hind margin of elytra reddish-brown
Number of spots 0
Pronotum Black (with reddish-brown margin in male)
Leg colour Reddish-brown
Other features Hairy; tip of abdomen reddish-brown; head reddish-brown (male); black (female)

Food
Scymnus auritus is a predatory ladybird that feeds on Phylloxera (aphid-like insects). May act as a control for Grape Phylloxera, which can be a serious pest of vineyards and almost destroyed the wine industry in France after it was accidentally introduced in 1860.

Habitat
Scymnus auritus is a species of deciduous woodland and other habitats with oak trees. It is usually found on oak, and occasionally on other deciduous trees, such as sallow, Sycamore and Lime. There are just a few records of the species on low herbaceous plants.

S. auritus overwinters in bark crevices and under the bark of oak trees.

Suggested survey method
Tree beating, especially oaks.

Range
Depending on prey outbreaks, the distribution and abundance may vary substantially from year to year. Not recently recorded in Scotland.

National conservation status
Very local.

Distribution trend (1995–2015)
Decreasing.

life size

adult

Pine Scymnus

Scymnus suturalis
Thunberg, 1795

Elytra are a rich brown colour, with a characteristic dark line running down the centre line. One of the more common *Scymnus* species.

Identification (adult)
Length 1.5–2mm
Background colour Brown
Pattern Dark brown to black along centre line, especially at the top, thus sometimes forming a dark T-shape
Number of spots 0
Pronotum Black with brown tips, notably narrow
Leg colour Brown
Other features Hairy; head black

Food
Scymnus suturalis is a predatory ladybird that feeds on aphids and adelgids.

Habitat
Scymnus suturalis is a specialist of coniferous woodlands and other habitats with coniferous trees. It is found on needled conifers, especially Scots Pine and Douglas Fir. There are also records from larch, Juniper, and Western Hemlock-spruce, but it has only rarely been recorded on broad-leaved trees such as oak, sallow and birch.

S. suturalis overwinters in sheltered positions on needled conifers, especially using the buds on Scots Pine. It may also overwinter in moss.

Suggested survey method
Tree beating of pines.

Range
Broad distribution, with records scattered throughout Britain and Ireland.

National conservation status
Very local.

Distribution trend (1995–2015)
Stable.

139

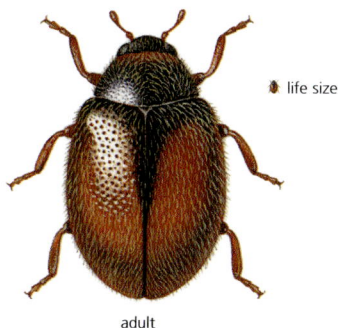

● life size

adult

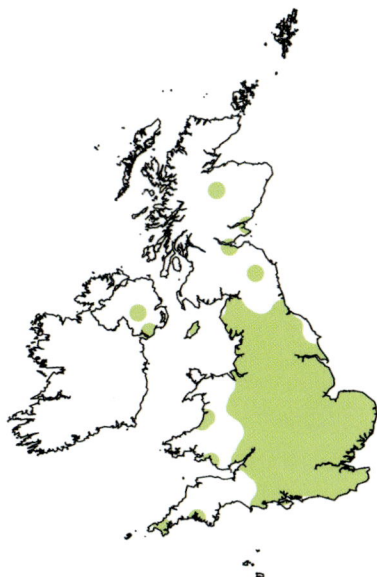

Heath Scymnus

Scymnus femoralis
(Gyllenhal, 1827)

Small, black, hairy ladybird with reddish-brown appendages. Very similar to *S. schmidti* but more convex and a little smaller. A relatively uncommon species with few recent records, but its small size and preference for being low down on vegetation may partly account for this.

140

Identification (adult)
Length 1.8–2.2mm
Background colour Black
Pattern None, or paler strip at rear of elytra
Number of spots 0
Pronotum Reddish-brown with black patch at base (male); black (female)
Leg colour Reddish-brown
Other features Hairy; reddish-brown antennae and mouthparts; head reddish-brown (male); black (female)

Food
Scymnus femoralis is a predatory ladybird that feeds on aphids.

Habitat
Scymnus femoralis is a species of heathland and other dry habitats, particularly on chalky or sandy soils. It is found low down on low-growing vegetation and occasionally on deciduous trees such as oaks.

S. femoralis overwinters in low herbage and moss, though there are few winter records of the species.

Suggested survey method
Sweep netting or visual searching of appropriate vegetation close to ground level.

Range
Limited distribution in England and Wales, but likely to be under-recorded.

National conservation status
Very local.

Distribution trend (1995–2015)
Insufficient data.

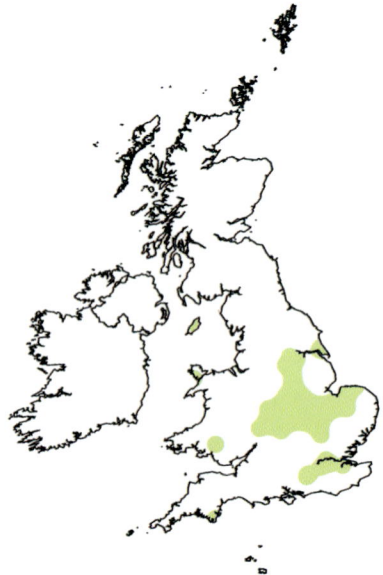

life size

form

adult

Angle-spotted Scymnus

Scymnus frontalis
(Fabricius, 1787)

The largest of the *Scymnus* species in Britain, and one of the commonest, at least in southeast and central England. The presence of an elongate red spot on each elytron helps to make *S. frontalis* one of the more recognisable of our small ladybirds.

Identification (adult)
Length 2.6–3.2mm
Background colour Black
Pattern Elongate red spots towards front of elytra
Number of spots 2
Pronotum Black, with brown margins in male only
Leg colour Brown
Other features Hairy; rather elongate; pointed abdomen; head brown (male), black (female); both with brown mouthparts

Food
Scymnus frontalis is a predatory ladybird that feeds on aphids.

Habitat
Scymnus frontalis is a species of heathland and other dry habitats, particularly on chalky or sandy soils, including dunes. It is found on low-growing vegetation including thistles and grasses, and occasionally in moss. It has rarely been recorded on woody species, including Gorse and pines.

S. frontalis overwinters in leaf litter, grass tussocks or the rosettes of perennial plants.

141

Suggested survey method
Sweep netting.

Range
Widely distributed in England and Wales. Preference for dry soils and bare ground, and this is reflected in the many coastal records.

National conservation status
Very local.

Distribution trend (1995–2015)
Stable.

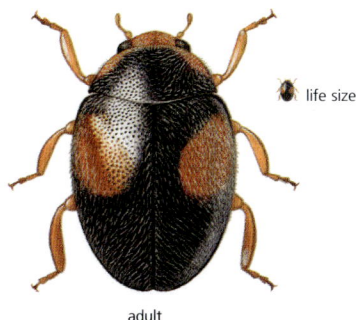

life size

adult

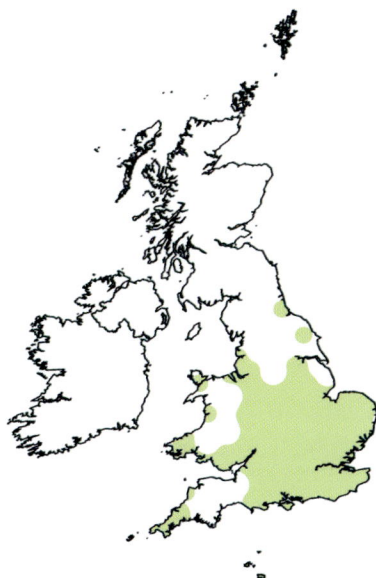

Red-flanked Scymnus

Scymnus interruptus
(Goeze, 1777)

Usually has a prominent triangular-shaped red spot on each elytron. *Scymnus interruptus* is a relatively common species in parts of continental Europe, including western France, where it occurs in diverse habitats including gardens and houses and, in Portugal, in citrus groves.

142

Identification (adult)
Length 1.5–2.2mm
Background colour Black
Pattern Red spots (roughly triangular in shape) at the front edges of the elytra
Number of spots 2
Pronotum Black
Leg colour Brown
Other features Hairy; head black

Food
Scymnus interruptus is a predatory ladybird that feeds on mealybugs and armoured scale insects.

Habitat
Scymnus interruptus may be found in various habitats in Britain, including gardens, where it is likely to be on hedges, shrubs and trees. It seems to favour Ivy. It has also been recorded on a range of woody species, including viburnum, privet, Firethorn and spindle.

There are very few winter records of *S. interruptus*, so its overwintering preferences are largely unknown. The few records available suggest that the species overwinters in woody species such as privet.

Suggested survey method
Beating of appropriate trees and shrubs.

Range
With the exception of a single record from 1986, all records are from 1996 or later. Some records are coastal and all but one come from southeast England. This may suggest that previously S. *interruptus* was an occasional migrant from Europe, rather than established in Britain. However, there have been increasing numbers of records in recent years and the species appears to be established, so perhaps it is benefiting from climate warming.

National conservation status
Very local.

Distribution trend (1995–2015)
Stable.

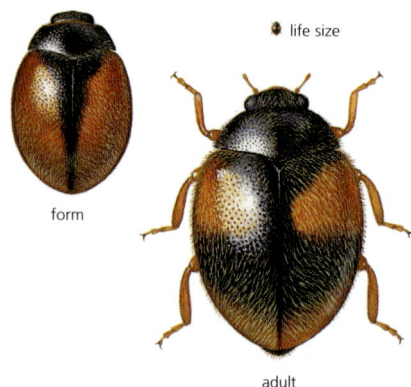

life size

form

adult

Black Scymnus

Scymnus nigrinus
Kugelann, 1794

The only species of *Scymnus* that is entirely black. A relatively specialist ladybird, reported in Surrey almost exclusively from young pine trees. However, there are very few recent records of the species.

Identification (adult)
Length 2–2.8mm
Background colour Black
Pattern None
Number of spots 0
Pronotum Black
Leg colour Black with pale to dark brown tarsi
Other features Hairy; head black

Food
Scymnus nigrinus is a predatory ladybird that feeds on aphids and adelgids.

Habitat
Scymnus nigrinus is a species of coniferous woodland and other habitats with coniferous trees. It lives on needled conifers, especially Scots Pine. It may also be found on Gorse and larch, and there is a single record of the species on oak.

 S. nigrinus overwinters on needled conifers, especially Scots Pine – usually in bark crevices, in sheltered positions around the buds and cones, or in the cones themselves.

Suggested survey method
Tree beating.

143

Range
Few recent records, but previously scattered records throughout Britain, and two records from the early 1900s from Ireland.

National conservation status
Very local.

Distribution trend (1995–2015)
Insufficient data.

adult

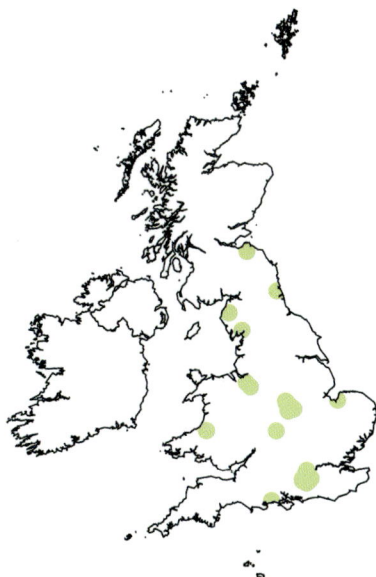

life size

Schmidt's Scymnus

Scymnus schmidti
Fürsch, 1958

144

Similar to, but rather larger and less convex than *Scymnus femoralis*, which is found in similar dry habitats.

Identification (adult)
Length 2.4–2.6mm
Background colour Black
Pattern None, or thin reddish margin at rear
Number of spots 0
Pronotum Black, with reddish anterior margins in male
Leg colour Brown (male); black with brown tibiae (female)
Other features Hairy; head reddish (male); black (female)

Food
Scymnus schmidti is a predatory ladybird that feeds on aphids.

Habitat
Scymnus schmidti is a species of dunes, dry grassland (including chalk grassland) and heathland. It is generally found low down in low-growing vegetation, including moss. Records from trees are rare, though it has been recorded on oak.

   S. schmidti overwinters in low herbage and in leaf litter.

Suggested survey method
Sweep netting or visual searching of appropriate vegetation close to ground level.

Range
Outside of southern and central England, records are sparse and nearly all coastal.

National conservation status
Very local.

Distribution trend (1995–2015)
Insufficient data.

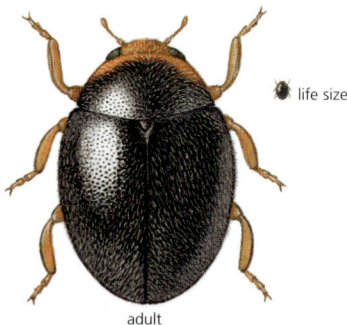

adult

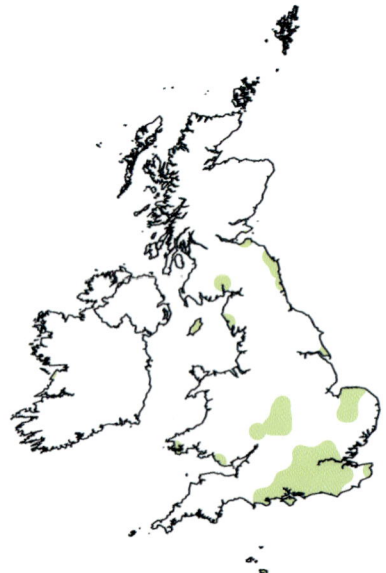

life size

Dot Ladybird

Stethorus pusillus
(Herbst, 1797)

One of Britain's tiniest ladybirds, with a varied habitat that probably reflects that of its prey, spider mites, some of which are serious plant and fruit pests. *Stethorus pusillus* can be abundant and acts as a good pest controller, for example in orchards, eating up to 150 spider mites each day.

gardens, hedgerows and grasslands, usually in association with its spider mite prey. It has been recorded on various deciduous trees (especially fruit trees and perhaps Buckthorn), Gorse and privet.

S. pusillus overwinters in bark crevices and under bark, choosing sheltered positions on deciduous trees, and possibly in leaf litter.

145

Identification (adult)
Length 1.3–1.5mm
Background colour Black
Pattern None
Number of spots 0
Pronotum Black
Leg colour Black and yellowish-brown (at least tibiae yellowish-brown)
Other features Antennae and mouthparts yellowish-brown; hairs towards tip of elytra lie parallel to suture (they diverge at an angle in *Scymnus* spp.); head black with yellowish-brown mouthparts

Food
Stethorus pusillus is a predatory ladybird that feeds on spider mites and small aphids.

Habitat
Stethorus pusillus occurs in various habitats including deciduous woodlands, orchards,

Suggested survey method
Tree beating.

Range
Under-recorded, owing to its minute size. Does not occur in Scotland or Ireland.

National conservation status
Very local.

Distribution trend (1995–2015)
Decreasing.

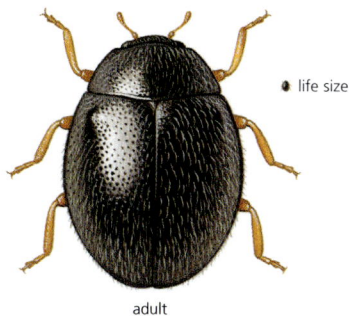

adult

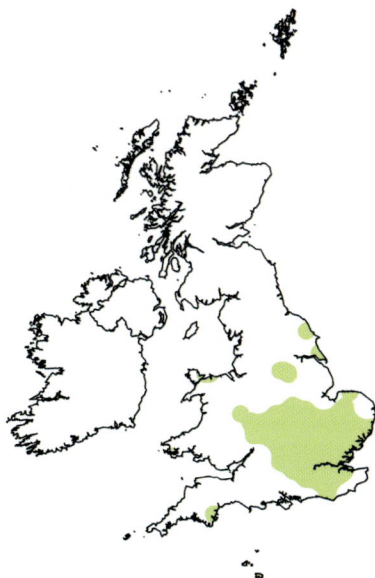

● life size

Similar species

This section provides help on how to distinguish ladybirds from similar-looking insects in other groups, and on some of the ladybird species that are tricky to tell apart from each other.

Non-ladybirds

There are a few other insects (mostly but not exclusively beetles) that may be confused with ladybirds, at least superficially. Some red and black true bugs (such as shieldbugs) can look rather like ladybirds, and the juveniles (nymphs) may be rather similar in shape. On closer inspection the folded wing-cases (hemelytra) of these insects are more textured and membranous compared to the smooth hard wing-cases (elytra) of ladybirds. Although there is variability among ladybird species, they generally have shorter antennae than shieldbugs or leaf beetles. Similarly, ladybird antennae are simple thread-like structures with a discrete club at the end, so any insect with elaborate antennae (feathered, comb-like or with plates, for example) is not a ladybird.

▲ Poplar Beetle *Chrysomela populi* (Coleoptera: Chrysomelidae).

▲ Scarlet Shieldbug *Eurydema dominulus* (Hemiptera: Pentatomidae).

▲ Rosemary Beetle *Chrysolina americana* (Coleoptera: Chrysomelidae).

▲ Gorse Shieldbug *Piezodorus lituratus* nymph (Hemiptera: Pentatomidae).

▲ False Ladybird *Endomychus coccineus* (Coleoptera: Endomychidae).

Ladybirds

On the pages that follow, pages 147–151, we have brought together 39 illustrations used elsewhere in the book to show species of ladybird that are easily confused, either as adults or as larvae. Please note that on the following pages, species are not shown to scale.

ADULTS

Black with four red spots

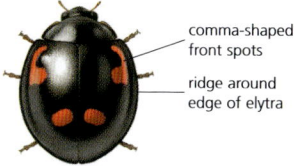

comma-shaped
front spots

ridge around
edge of elytra

Pine Ladybird
p.72

prominent
white 'false
eye'
markings

ridge or keel
at rear of
elytra

Harlequin Ladybird
f. *spectabilis* p.106

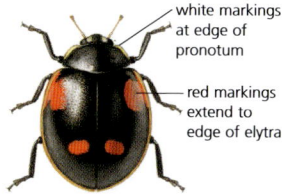

white markings
at edge of
pronotum

red markings
extend to
edge of elytra

2-spot Ladybird
f. *quadrimaculata* p.90

Black with two red spots

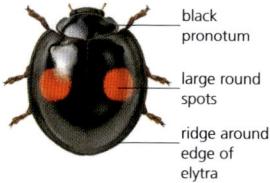

black
pronotum

large round
spots

ridge around
edge of
elytra

Kidney-spot Ladybird
p.70

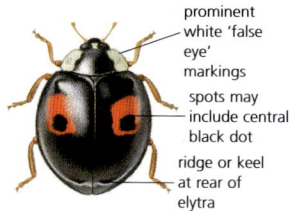

prominent
white 'false
eye'
markings

spots may
include central
black dot

ridge or keel
at rear of
elytra

Harlequin Ladybird
f. *conspicua* p.106

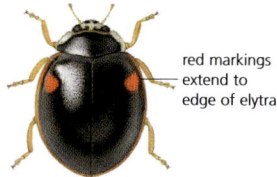

red markings
extend to
edge of elytra

10-spot Ladybird
f. *bimaculata* p.93

Orange-red with many spots

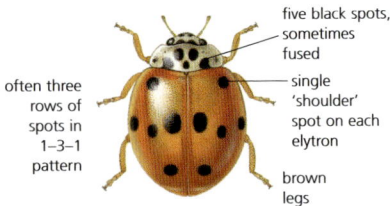

five black spots, sometimes fused

often three rows of spots in 1–3–1 pattern

single 'shoulder' spot on each elytron

brown legs

10-spot Ladybird
f. *decempunctata* p.93

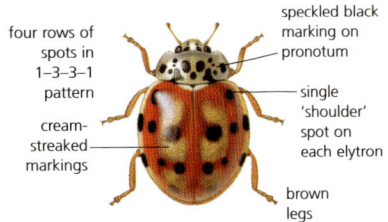

four rows of spots in 1–3–3–1 pattern

speckled black marking on pronotum

cream-streaked markings

single 'shoulder' spot on each elytron

brown legs

Cream-streaked Ladybird ('16-spotted' form)
p.109

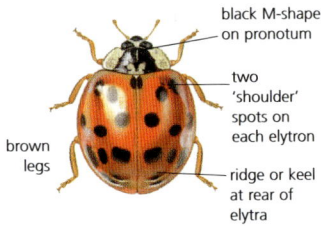

black M-shape on pronotum

two 'shoulder' spots on each elytron

brown legs

ridge or keel at rear of elytra

Harlequin Ladybird f. *succinea*
p.106

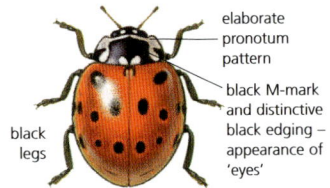

elaborate pronotum pattern

black M-mark and distinctive black edging – appearance of 'eyes'

black legs

Eyed Ladybird (without rings)
p.114

Red with seven black spots

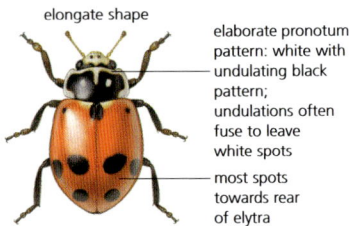

elongate shape

elaborate pronotum pattern: white with undulating black pattern; undulations often fuse to leave white spots

most spots towards rear of elytra

Adonis Ladybird
p.86

domed shape

four white triangular markings on underside (not shown)

simple white 'false eye' markings

large central spots

Scarce 7-spot Ladybird
p.98

two white triangular markings on underside (not shown)

simple white 'false eye' markings

small central spots

7-spot Ladybird
p.102

Small and yellow with black markings

four discrete black spots in a semi-circle, black triangle at the mid-base; pronotum yellow or white

bright lemon-yellow elytra

20–22 small black spots

22-spot Ladybird
p.78

pale elytra

fused row of spots forming line

16-spot Ladybird
p.82

solid black crown-shaped pattern

rectangular black markings

spots often joined to form anchor shape

14-spot Ladybird
p.112

large pale spots on dark elytra (rather than vice versa)

10-spot Ladybird
f. *decempustulata* p.93

Orange-brown with white spots

orange legs

translucent edge of pronotum

spots in lines down elytra

Orange Ladybird
p.76

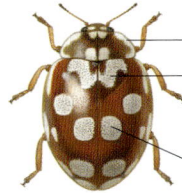

elaborate pronotum pattern

L-shaped cream marking around scutellum

irregular cream spots

18-spot Ladybird
p.116

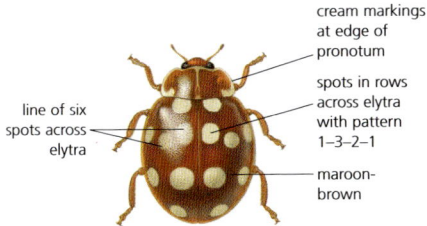

cream markings at edge of pronotum

spots in rows across elytra with pattern 1–3–2–1

maroon-brown

line of six spots across elytra

Cream-spot Ladybird
p.118

LARVAE

Dark grey with lateral orange stripe

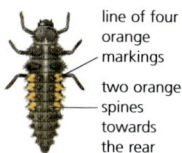

line of four orange markings

L-shaped line of five orange markings

two orange spines towards the rear

four orange spines towards rear

Cream-streaked Ladybird p.109

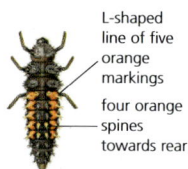

Harlequin Ladybird p.106

Grey with one central and two lateral orange-yellow spots

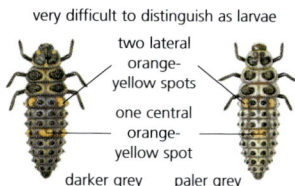

very difficult to distinguish as larvae

two lateral orange-yellow spots

one central orange-yellow spot

darker grey

paler grey

2-spot Ladybird p.90

10-spot Ladybird p.93

Dark with long bristles

pale stripe across abdomen

plain reddish-brown

Heather Ladybird p.68

Kidney-spot Ladybird p.70

Pale with profuse branched spikes

pale spikes on creamy-yellow/greenish

dark grey spikes on pale yellow

24-spot Ladybird p.124

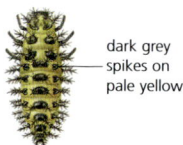

Bryony Ladybird p.122

Yellow with dark bristles

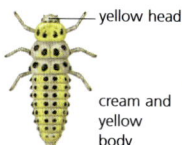

yellow head

dark head

cream and yellow body

yellow body

Orange Ladybird p.76

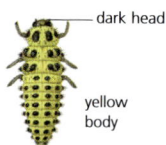

22-spot Ladybird p.78

Grey with lateral orange-yellow markings

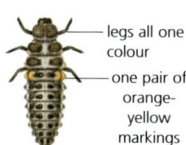

legs all one colour

legs bicoloured

one pair of orange-yellow markings

Larch Ladybird p.88

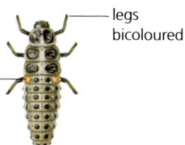

18-spot Ladybird p.116

Dark grey with orange-yellow spots

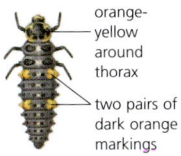

orange-yellow around thorax

two pairs of dark orange markings

7-spot Ladybird
p.102

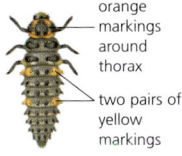

orange markings around thorax

two pairs of yellow markings

Scarce 7-spot Ladybird
p.98

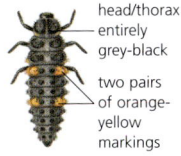

head/thorax entirely grey-black

two pairs of orange-yellow markings

11-spot Ladybird
p.104

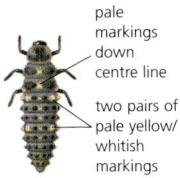

pale markings down centre line

two pairs of pale yellow/whitish markings

Hieroglyphic Ladybird
p.96

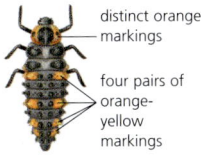

distinct orange markings

four pairs of orange-yellow markings

5-spot Ladybird
p.100

Heavily patterned with pale markings

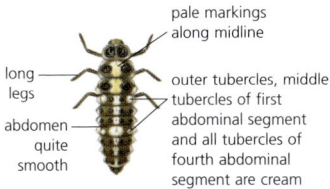

pale markings along midline

long legs

abdomen quite smooth

outer tubercles, middle tubercles of first abdominal segment and all tubercles of fourth abdominal segment are cream

14-spot Ladybird
p.112

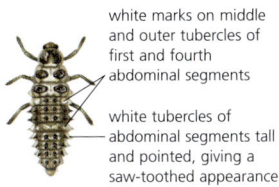

white marks on middle and outer tubercles of first and fourth abdominal segments

white tubercles of abdominal segments tall and pointed, giving a saw-toothed appearance

Cream-spot Ladybird
p.118

Potential new species

With suitable weather conditions, ladybirds are able to disperse long distances. There are far more ladybird species in mainland Europe than in Britain and Ireland; for example, Belgium has about 20 species that we lack. These two points together lead to the likelihood of new arrivals of ladybird species to our islands. Perhaps increasingly, global warming is making our climate more suitable for some species that are towards the edges of their ranges, so they may become established, at least in southeastern England if not elsewhere in Britain and Ireland. Here we briefly discuss some of these potential (and in two cases actual) new arrivals.

Oenopia conglobata

Oenopia conglobata is a widespread and relatively common ladybird species in parts of Europe. It is primarily a species of deciduous trees and occurs close to Britain, in coastal areas of Belgium and France. We now have a single British record of this rather attractive pink ladybird, which has wing-case (elytral) patterning that resembles musical notes. It was found in 2014, not in southeast England, but near Flamborough Head on the Yorkshire coast. How it came to be there is unknown, and it presumably arrived inadvertently with the aid of people. There are no other records of the species so this may be a one-off for now, but it is a likely candidate for further arrivals and potential establishment in the near future.

Cynegetis impunctata

Cynegetis impunctata is a small, hairy, red-brown ladybird that resembles an unspotted 24-spot Ladybird. Like the 24-spot, it is a grassland species. Unusually for ladybirds, *C. impunctata* lacks the ability to fly, thus restricting its dispersal potential. However, Majerus (1994) reported the species to be abundant around the French terminus of the Channel Tunnel, and argued that it was a strong candidate for immigration to Britain: it may arrive with people on trains or cars.

Calvia decemguttata

Calvia decemguttata was recorded in Ireland (Killarney) in 1927 and there is a single nineteenth-century record of it in western England (Majerus 1994). This species resembles its close relative the Cream-spot Ladybird and also looks rather similar to the Orange Ladybird. However, as suggested by

its scientific name, *C. decemguttata* generally has 10 spots, fewer than those of the other two ladybirds. The species usually lives on deciduous trees, where it feeds on aphids. It is fairly likely that *C. decemguttata* may make it over the English Channel again in due course.

Vibidia duodecimguttata

Vibidia duodecimguttata (sometimes known as the 12-spot Ladybird) is included in the British list of ladybird species (Duff 2018), and there are old records from England, Scotland, Northern Ireland and the Republic of Ireland. However, apart from several recent British records that require confirmation and a western Ireland record from 1973, the records date back to the early twentieth century or earlier. *V. duodecimguttata* is a common species in mainland Europe. It has most likely only ever been an occasional migrant to Britain and Ireland (Majerus 1994), but it could become established.

V. duodecimguttata looks like a small version of the Orange Ladybird, but generally has two fewer spots. Like the Orange Ladybird, it feeds on mildew on deciduous trees.

Rhyzobius forestieri

The most recent ladybird species to arrive in Britain is *Rhyzobius forestieri*. First discovered in London in 2014, *R. forestieri* is a hairy inconspicuous ladybird that is native to Australia. It is larger than most of the other small ladybirds, and approaches the size of a 24-spot Ladybird. *R. forestieri* is black on top, but when viewed from underneath the rear segments of the abdomen are a bright orange colour, making it quite distinctive. There have since been other records of the species in London and Kent, and in 2017 it was found for the first time in both Cambridgeshire and Essex. This species has recently been added to the list of established ladybird species in Britain.

Useful resources

Further reading

Bradbury, K. (2013) *The Wildlife Gardener.* Kyle Books, London.

Comont, R. (in press) *RSPB Spotlight Ladybirds.* Bloomsbury Publishing, London.

Duff, A.G. (ed.) (2018) *Checklist of Beetles of the British Isles.* 3rd edition. Pemberley Books, Iver.

Hawkins, R.D. (2000) *Ladybirds of Surrey.* Surrey Wildlife Trust, Woking.

JNCC (2018) Conservation designations for UK taxa: http://jncc.defra.gov.uk/page-3408.

Majerus M.E.N. (1994) *Ladybirds.* New Naturalist 81. Collins, London.

Majerus, M.E.N., Roy, H.E. and Brown, P.M.J. (2016) *A Natural History of Ladybird Beetles.* Cambridge University Press, Cambridge.

Nedvěd, O. (2015) *Ladybird Beetles (Coccinellidae) of Central Europe.* Academia, Czech Republic.

Roy, H.E., Brown, P.M.J., Comont, R.F., Poland, R.L. and Sloggett, J.J. (2013) *Ladybirds.* Naturalists' Handbook 10. Pelagic Publishing, Exeter.

Roy, H.E., Brown, P.M.J., Frost, R. and Poland, R.L. (2011) *Ladybirds (Coccinellidae) of Britain and Ireland.* Biological Records Centre, Centre for Ecology & Hydrology, Wallingford.

Websites, apps and social media

There are a number of websites and smartphone apps for online recording, which also provide information:

All Ireland Ladybird Survey: www.biology.ie/home.php?m=ladybirds2

iRecord app: https://irecord.org.uk/app

iRecord Ladybirds app: www.brc.ac.uk/article/irecord-ladybirds-mobile-app

Ladybirds of Ireland: www.habitas.org.uk/ladybirds

Reports on the distribution of species can be found through the maps on the NBN Atlas website or through the UK Ladybird Survey and UK Beetle Recording websites:

NBN Atlas: https://nbnatlas.org

UK Ladybird Survey: www.ladybird-survey.org – the website of the national recording scheme for ladybirds in the UK. Information about the identification and ecology of ladybirds, with links for recording them (in iRecord – see above). Regular feedback and news updates on Twitter: @UKLadybirds

UK Beetle Recording: www.coleoptera.org.uk

Record centres

Biological Records Centre, Centre for Ecology & Hydrology, Wallingford, Oxfordshire, UK: www.brc.ac.uk

National Biodiversity Data Centre, Beechfield House, Waterford Institute of Technology West Campus, Carriganore, Co. Waterford, Ireland: www.biodiversityireland.ie

Local Environmental Record Centres: www.alerc.org.uk

Glossary

List of explanatory terms and synonyms, for use with this guide and in additional reading.

Abdomen Third section of an insect, containing the digestive tract and reproductive organs. Comprises a series of concave upper integumental plates (**tergites**) and convex lower integumental plates (**sternites**), held together by a tough flexible membrane.

Abdominal tergites Sclerotised (hardened) integumental (exoskeleton) plates on the upper part of the abdomen.

Adelgids Woolly conifer aphids; insect family Adelgidae within the order Hemiptera (true bugs).

Aestivation Change in adult activity in order to survive summer, e.g. moving to cooler areas.

Aleyrodids Whiteflies; insect family Aleyrodidae within the order Hemiptera (true bugs).

Alkaloids Repellent chemicals contained within ladybird blood and secreted through reflex bleeding as a defence.

Anal cremaster Anal pore located at the tip of the last abdominal segment of an insect.

Anterior Front end of an insect.

Aphidophagous Feeds on aphids.

Aphids Greenfly or blackfly; insect family Aphididae within the order Hemiptera (true bugs).

Aposematism Bold contrasting patterns representing warning colouration.

Bivoltine See **Voltinism**.

Chorion Outer shell of an insect egg.

Chrysomelids Seed or leaf beetles; insect family Chrysomelidae.

Coccidophagous Feeds on coccids.

Coccids Scale insects; insect superfamily Coccoidea within the order Hemiptera (true bugs).

Congeneric Species within the same genus; e.g. 2-spot (*Adalia bipunctata*) and 10-spot (*Adalia decempunctata*).

Conspicuous ladybirds Large, brightly coloured ladybirds.

Coxa Segment of leg that connects it to the thorax ('hip'). Plural: coxae.

Diapause Period during which growth or development is delayed and metabolic activity is reduced, often in response to adverse environmental conditions.

Diaspidids Armoured scale insects; insect family Diaspididae within the order Hemiptera (bugs).

Dormancy Inactive state that promotes the survival of insects during adverse environmental conditions.

Dorsal Upper surface. Antonym: ventral.

Dorsoventrally flattened Flattened from top to bottom (squashed in appearance).

Eclosion Emergence of the adult stage from the pupa.

Elytron Hardened forewing. Plural: elytra.

Exoskeleton External skeleton (integument) that supports and protects an insect's body.

Exudate Fluid oozed during reflex bleeding.

Exuvia Shed larval skin. Plural: exuviae.

Family Taxonomic rank, below order and above genus. Standard nomenclature rules end family zoological names with '-idae'.

Femur Long and relatively thick leg segment above the 'knee' ('thigh'). Plural: femora.

Form The name given to a morph or colour form of a particular species. Abbreviation: f.

Frons Dorsal part of face, i.e. the 'nose' or 'forehead'.

Genus A taxonomic category in which species are classified, e.g. *Coccinella septempunctata* belongs to the genus *Coccinella*. Plural: genera.

Haemolymph Insect blood.

Head First section of an insect, where the antennae, compound eyes and mouthparts are located; the major centre of the nervous system.

Holometabolous Insects that undergo complete metamorphosis, including four stages: egg, larva, pupa, adult. Contrast with **hemimetabolous** insects, which undergo incomplete metamorphosis and have juveniles (nymphs) that look like miniature versions of the adult forms.

Inconspicuous ladybirds Small, indistinct and slightly hairy species that may not be easily recognised as ladybirds.

Instar Phase between two periods of moulting in the development of an insect larva.

Integument External skeleton (exoskeleton) that supports and protects an insect's body.

Labium Lip-like structure covering the mandibles on the underside of the head, i.e. the 'lower lip'.

Labrum Lower portion of face, a lip-like structure shielding the mandibles from the front, i.e. the 'upper lip'.

Lateral Along the sides.

Lateral carina Ridge along each side of abdomen, separating upper- and underside.

Longitudinal Running lengthwise.

Melanic Dark colour forms of adult ladybird.

Mesothorax Middle of the three segments in the thorax of an insect, bearing the second pair of legs and forewings. It comprises sclerites (exoskeletal plates): **mesonotum** (dorsal), **mesosternum** (ventral), and **mesopleuron** (lateral) on each side.

Metatarsus First section of the tarsus (foot).

Methylalkylpyrazine Repellent chemical contained within ladybird blood and secreted through reflex bleeding as defence.

Mildew Mould (fungus) often seen growing on plants.

Multivoltine See **Voltinism**.

Mycophagous Feeds on fungi (mould or mildew).

Myrmecophile Animal that strongly associates with ants.

Ocellus Light-sensitive organ appearing as a small single-lens eye; three of these lie between and/or in front of the compound eyes. Plural: ocelli.

Ommatidia Optical units that make up the compound eye of an insect.

Oviposition Egg laying.

Parasitoid An insect whose larvae live as parasites which eventually kill their hosts (parasites do not usually kill the host).

Pentatomids Shieldbugs; insect family Pentatomidae within the order Hemiptera (true bugs).

Pheromone A chemical released by an animal that influences the physiology or behaviour of other members of the same species.

Phylloxera Small aphid-like insects; insect family Phylloxeridae within the order Hemiptera (true bugs).

Phylogenetic Evolutionary relationships among species or group of organisms.

Phytophagous Feeds on plants.

Polymorphism Occurrence of two or more different forms; for ladybirds refers especially to substantial variation in elytral (wing-case) colour patterns.

Posterior Rear end of an insect.

Pronotum Shield-like plate covering the top of the prothorax; the colour and pattern of the pronotum is diagnostic in many adult ladybirds. Plural: pronota.

Prosternum Ventral sclerite of the prothorax of an insect.

Prothorax Small separate portion of thorax behind head, bearing forelegs.

Pseudococcids Mealybugs; insect family Pseudococcidae within the order Hemiptera (true bugs).

Psyllids Jumping plant lice; insect family Psyllidae within the order Hemiptera (true bugs).

Quiescence State of dormancy from which the individual can emerge at any time should conditions become favourable again.

Reflex bleeding Defensive reaction in which reflex blood is exuded.

Reflex blood Insect blood (haemolymph) secreted through the process of **reflex bleeding**; commonly observed as yellow droplets from the leg joints in adults, and pores in the dorsal surface of larvae and pupae.

Sclerite Plates forming the exoskeleton.

Scutellum Posterior portion of the mesonotum of an insect thorax – a small shield-like plate at the front of the wing-cases.

Senescence Biological ageing.

Species Unit of biological classification and a taxonomic rank; largest group of organisms in which two individuals can produce fertile offspring, typically by sexual reproduction. Abbreviation: sp., plural spp.

Subfamily Intermediate taxonomic rank, below family but more inclusive than genus. Standard nomenclature rules end subfamily zoological names with '–inae'. The only subfamily of the family Coccinellidae represented in Britain and Ireland is the Coccinellinae.

Subspecies Category in which discrete regional varieties of a species are classified. Abbreviation: ssp.

Tarsal claws Paired hooks at tips of legs.

Tarsus Group of small terminal leg segments ('foot'). Plural: tarsi.

Tetranychids Spider mites; arachnid family Tetranychidae within the subclass Acari and order Trombidiformes (mites).

Thorax Second (middle) section of insect body, bearing legs and wings.

Tibia Long and relatively thin leg segment below the 'knee' (the 'shin'). Plural: tibiae.

Transverse Extending across.

Tribe Taxonomic rank above genus, but below family and subfamily. Standard nomenclature rules end zoological tribe names with '–ini'.

Trochanter Second of five sections (of a coxa, trochanter, femur, tibia, tarsus) of an insect's leg.

Tubercle Round nodule, small eminence or warty outgrowth.

Univoltine See **Voltinism**.

Ventral Lower surface. Antonym: dorsal.

Voltinism Number of generations completed in one year. For example, Seven-spot Ladybirds are **univoltine** (one generation completed in one year) and Harlequin Ladybirds are **bivoltine** (two generations completed in one year) or sometimes **multivoltine** (many generations completed in one year).

Plant names

Alder	*Alnus glutinosa*
Angelica	*Angelica sylvestris*
Apple	*Malus* spp.
Ash	*Fraxinus excelsior*
Beech	*Fagus sylvatica*
Birch	*Betula* spp.
Bitter-cress	*Cardamine hirsuta*
Black Pine	*Pinus nigra*
Blackcurrant	*Rubus nigrum*
Blackthorn	*Prunus spinosa*
Bracken	*Pteridium aquilinum*
Bramble	*Rubus fruticosus*
Broom	*Cytisus scoparius*
Buckthorn	*Rhamnus cathartica*
Buddleja	*Buddleja davidii*
Burdock	*Arctium* spp.
Buttercup	*Ranunculus* spp.
Camellia	*Camellia japonica*
Cherry	*Prunus* spp.
Corsican Pine	*Pinus nigra* ssp. *laricio*
Cow Parsley	*Anthriscus sylvestris*
Creeping Thistle	*Cirsium arvense*
Cypress	*Cupressus* spp.
Dandelion	*Taraxacum officinale*
Dead-nettle	*Lamium album,*
	L. purpureum
Dock	*Rumex* spp.
Dogwood	*Cornus sanguinea*
Douglas Fir	*Pseudotsuga menziesii*
Elder	*Sambucus nigra*
Elm	*Ulmus* spp.
False Oat-grass	*Arrhenatherum elatius*
Fat-hen	*Chenopodium album*
Field Bean	*Vicia faba*
Field Maple	*Acer campestre*
Fir	*Abies* spp.
Firethorn	*Pyracantha coccinea*
Foxglove	*Digitalis purpurea*
Gorse	*Ulex europaeus*
Hawthorn	*Crataegus monogyna*
Hazel	*Corylus avellana*
Heather	*Calluna vulgaris, Erica* spp.
Hebe	*Veronica* subgenus
	Pseudoveronica
Hogweed	*Heracleum sphondylium*
Holly	*Ilex aquifolium*
Honeysuckle	*Lonicera periclymenum*
Hornbeam	*Carpinus betulus*
Horse-chestnut	*Aesculus hippocastanum*
Ivy	*Hedera helix*
Juniper	*Juniperus communis*
Knapweed	*Centaurea* spp.
Larch	*Larix* spp.
Lavender	*Lavandula angustifolia*

Leyland Cypress	*Cupressus macrocarpa* ×
	Xanthocyparis
	nootkatensis (×
	Cuprocyparis leylandii)
Lime	*Tilia* × *europaea*
Lucerne	*Medicago sativa*
Maple	*Acer* spp.
Melon	*Cucumis melo*
Monterey Cypress	*Cupressus macrocarpa*
Mugwort	*Artemisia vulgaris*
Nettle	*Urtica dioica*
Norway Spruce	*Picea abies*
Oak	*Quercus* spp.
Oleaster	*Elaeagnus* spp.
Peony	*Paeonia officinalis*
Pine	*Pinus* spp.
Plane	*Platanus* × *hispanica*
Poplar	*Populus* spp.
Privet	*Ligustrum* spp.
Ragwort	*Senecio* spp.
Red Campion	*Silene dioica*
Reed	*Phragmites australis*
Reed Sweet-grass	*Glyceria maxima*
Reedmace	*Typha latifolia*
Rhododendron	*Rhododendron ponticum*
Rose	*Rosa* spp.
Rosebay Willowherb	*Chamerion angustifolium*
Rush	*Juncus* spp.
Salad Burnet	*Poterium sanguisorba*
Sallow	*Salix caprea, S. cinerea*
Scots Pine	*Pinus sylvestris*
Sea-buckthorn	*Hippophae rhamnoides*
Sea-lavender	*Limonium* spp.
Sea Radish	*Raphanus raphanistrum*
	ssp. *maritimus*
Sessile Oak	*Quercus petraea*
Spindle	*Euonymus* spp.
Spruce	*Picea* spp.
Spurrey	*Spergula arvensis*
Sycamore	*Acer pseudoplatanus*
Tansy	*Tanacetum vulgare*
Teasel	*Dipsacus fullonum*
Thistle	*Cirsium* spp.
Viburnum	*Viburnum* spp.
Wallflower	*Erysimum cheiri*
Western	*Tsuga heterophylla*
Hemlock-spruce	
White Bryony	*Bryonia dioica*
Wild Carrot	*Daucus carota*
Wild Parsnip	*Pastinaca sativa*
Willow	*Salix fragilis, S. alba*
Yarrow	*Achillea millefolium*
Yellow Iris	*Iris pseudacorus*
Yew	*Taxus baccata*

Animal names

List of names of animals other than ladybirds that appear throughout the text.

Adelgids	Adelgidae
Aphids	Aphididae
Armoured scale insects	Diaspididae
Black Garden Ant	*Lasius niger*
Blue Tit	*Cyanistes caeruleus*
Braconid wasp	*Dinocampus coccinellae*
Encyrtid wasp	*Homalolytus* spp.
Eulophid wasp	*Aprostocetus* spp.
Eulophid wasp	*Oomyzus* spp.
False Ladybird	*Endomychus coccineus*
Flea beetle	*Altica* spp.
Flea beetle	*Galerucella* spp.
Glanville Fritillary	*Melitaea cinxia*
Gorse Shieldbug	*Piezodorus lituratus*
Grape Phylloxera	*Daktulosphaira vitifoliae*
Heather Aphid	*Aphis callunae*
Heather Leaf Beetle	*Lochmaea suturalis*
Jumping plant lice	Psyllidae
Leaf beetles	Chrysomelidae
Mealy Plum Aphid	*Hyalopterus pruni*
Mealybug	*Pseudococcus* spp.
Mites	Acari
Northern Wood Ant	*Formica lugubris*
Oak Leaf Phylloxera	*Phylloxera glabra*
Phorid fly	*Phalacrotophora beroliensis*
Phorid fly	*Phalacrotophora fasciata*
Pine aphid	*Cinara* spp.
Poplar Beetle	*Chrysomela populi*
Red Wood Ant	*Formica rufa*
Rosemary Beetle	*Chrysolina americana*
Scale insects	Coccoidae
Scarlet Shieldbug	*Eurydema dominulus*
Spider mites	Tetranychidae
Tachinid fly	*Medina separata*
Whiteflies	Aleyrodidae
Woolly aphids	Aphididae (Eriosomatinae)

Photographic credits

Bloomsbury Publishing would like to thank the following for providing images and for permission to reproduce copyright material. While every effort has been made to trace and acknowledge all copyright holders, we would like to apologise for any errors or omissions and invite readers to inform us so that corrections can be made in any future editions of the book.

6 © Chris Doward; **7** both © Gilles San Martin; **11** top left, top right, bottom right © Gilles San Martin, centre right © Richard Musgrave, bottom left © Peter Brown; **14** left © Gilles San Martin, right © Peter Brown; **15** © Gilles San Martin; **16** both © Gilles San Martin; **17** © Gilles San Martin; **19** © Gilles San Martin; **20** © Anne Riley; **21** top © Helen Roy, centre © Nick Hollands, bottom © Geoff Foale; **23** © Gilles San Martin; **24** © Gilles San Martin; **25** © Gilles San Martin; **26** top © Ken Dolbear, bottom © Gilles San Martin; **27** © Gilles San Martin; **28** © Gilles San Martin; **29** © Trevor James; **31** both © Helen Roy; **32** top © Peter Brown, bottom © Helen Roy; **33** © Helen Roy; **35** top © Matylda Laurence/Shutterstock, centre © Ian_Stewart/Shutterstock, bottom © Helen Roy; **36** both © Helen Roy; **37** © Jonathan Mitchell Images/Shutterstock; **38** © David Peter Robinson/Shutterstock; **39** top © Erni/Shutterstock, bottom © Helen Roy; **40** © John van Breda; **41** © Peter Brown; **42** © Nick Hawkes/Shutterstock; **43** © Katy Roper; **44** © Trevor James; **45** top © Allen Paul Photography/Shutterstock, bottom © Jason Wells/Shutterstock; **46** © Nicola Pulham/Shutterstock; **47** © Derek Bateson; **49** © Sue Burton Photography Ltd/Shutterstock; **50** © Dave McAleavy/Shutterstock;

51 © Peter Brown; **52** © Katie Berry; **53** © Gill Weyman; **54** © Peter Brown; **55** © Gilles San Martin; **61** © Gilles San Martin; **68** © Gilles San Martin; **70** © Gilles San Martin; **72** © Gilles San Martin; **74** © Peter Brown; **76** © Gilles San Martin; **78** © Peter Brown; **80** © Gilles San Martin; **82** © Gilles San Martin; **84** © Gilles San Martin; **86** © Peter Brown; **88** © Gilles San Martin; **90** © Gilles San Martin; **92** top both © Mike Majerus, bottom © Gilles San Martin; **93** © Gilles San Martin; **95** © Sandy Rae; **96** © Maris Midgley; **98** © Peter Brown; **100** © Gilles San Martin; **102** © Gilles San Martin; **104** © Peter Brown; **106** © Gilles San Martin; **109** © Gilles San Martin; **110** © Gilles San Martin; **111** both © Gilles San Martin; **112** © Gilles San Martin; **114** © Peter Brown; **116** © Gilles San Martin; **118** © Gilles San Martin; **120** © Gilles San Martin; **122** © Gilles San Martin; **124** © Gilles San Martin; **146** top © Jiri Prochazka/Shutterstock, centre left © Mike Majerus, centre right © stefg1971/Shutterstock, bottom left © Sandra Standbridge/Shutterstock, bottom right © IanRedding/Shutterstock; **152** all © Peter Brown; **153** top © Peter Brown, bottom © Harry Taylor/Natural History Museum.

Index

Main entries are in **bold**.